Vertically-Oriented Graphene

Junhong Chen · Zheng Bo · Ganhua Lu

Vertically-Oriented Graphene

PECVD Synthesis and Applications

Junhong Chen
Department of Mechanical Engineering, Department of Materials Science and Engineering
University of Wisconsin-Milwaukee
Milwaukee
WI
USA

Ganhua Lu
Department of Mechanical Engineering
University of Alaska Anchorage
Anchorage
AK
USA

Zheng Bo
State Key Laboratory of Clean Energy Utilization, College of Energy Engineering
Zhejiang University
Hangzhou
Zhejiang
China

ISBN 978-3-319-38144-2 ISBN 978-3-319-15302-5 (eBook)
DOI 10.1007/978-3-319-15302-5

Springer Cham Heidelberg New York Dordrecht London

Softcover reprint of the hardcover 1st edition 2015

Printed on acid-free paper

Springer International Publishing AG Switzerland is part of Springer Science+Business Media (www.springer.com)

Preface

This book introduces the basic concepts, synthesis techniques, and applications of vertically-oriented graphene (VG), which has recently attracted growing interest for a wide range of applications due to its unique orientation, exposed sharp edges, non-stacking morphology, and large surface-to-volume ratio. The book summarizes the state-of-the-art research on the synthesis of vertically-oriented graphene nanosheets. Particularly, this book provides a detailed introduction to the plasma-assisted growth of vertically-oriented graphene toward massive industrial production. Emerging applications of vertically-oriented graphene such as biosensors and gas sensors, atmospheric nanoscale corona discharges, supercapacitors, lithium-ion batteries, fuel cells (catalyst support), and solar cells are discussed in this book. The intended readers of this book include upper level undergraduate students, graduate students, and material scientists and researchers.

Acknowledgments

We wish to express our deep gratitude to a number of agencies that have funded our research related to vertically-oriented graphene (VG). Specifically, Chen gratefully acknowledges the financial support from the US National Science Foundation (CMMI-0900509, ECCS-1001039, and IIP-1128158), the US Department of Energy (DE-EE0003208), and the Research Growth Initiative Program of the University of Wisconsin-Milwaukee (UWM). Bo acknowledges the financial support from the National Natural Science Foundation of China (No. 51306159). Lu thanks the University of Alaska Anchorage for an Innovate Award and Faculty Development Grants for financial support. We were inspired to prepare this book by some exciting studies we have carried out on VG in the past several years. To some extent, this book is a summary of those studies, particularly two review articles: "Plasma-Enhanced Chemical Vapor Deposition Synthesis of Vertically-Oriented Graphene Nanosheets," *Nanoscale* **5**(12), 5180–5204, 2013 (DOI: 10.1039/C3NR33449J); and "Emerging Energy and Environmental

Applications of Vertically-Oriented Graphenes," *Chemical Society Reviews*, 2015 (DOI: 10.1039/C4CS00352G). Our studies on VG could not have been successful without the contributions from many individuals. We thank graduate students, undergraduate students, and postdocs, present and former, in the Nanotechnology for Sustainable Energy and Environment Laboratory at UWM and in the Energy Storage, Nanocatalysis, Plasma Technology Laboratory at Zhejiang University for their well-done work on VG research, and collaborators around the world for fruitful discussions and insightful suggestions. Particularly, we thank Prof. Rodney Ruoff at Ulsan National Institute of Science and Technology (Korea), Prof. Kostya (Ken) Ostrikov at the Commonwealth Scientific and Industrial Research Organisation (Australia), and Dr. Yong Yang at Huazhong University of Science and Technology (China) for productive collaboration. We thank Dr. Shun Mao for helping with the drafting of Chap. 6 and the revision of the entire manuscript.

January 2015

Junhong Chen
Zheng Bo
Ganhua Lu

Contents

1 Introduction ... 1
1.1 Versatility of Carbon ... 1
1.2 Graphene: A Two-Dimensional Carbon Nanostructure ... 2
1.3 Vertically-Oriented Graphene and Its Growth ... 2
1.4 Objectives and Organization of the Book ... 4
1.5 Summary ... 5
References ... 6

2 The Properties of Vertically-Oriented Graphene ... 11
2.1 General Characteristics of Graphene ... 11
2.2 Planar Graphene Versus Vertically-Oriented Graphene ... 13
2.3 Unique Properties of VG ... 14
2.4 Summary ... 16
References ... 16

3 PECVD Synthesis of Vertically-Oriented Graphene: Mechanism and Plasma Sources ... 19
3.1 VG Growth Mechanism by PECVD ... 20
3.2 Plasma Sources ... 23
3.2.1 Microwave Plasmas ... 26
3.2.2 Radio Frequency Plasmas ... 27
3.2.3 Direct Current Plasmas ... 28
3.3 Raman Spectroscopy of VG ... 29
3.4 Summary ... 31
References ... 32

4 PECVD Synthesis of Vertically-Oriented Graphene: Precursor and Temperature Effects ... 35
4.1 Precursors ... 35
4.1.1 Carbon Sources: Feedstock Gases ... 36
4.1.2 Amorphous Carbon Etchants ... 42

4.1.3 Argon 45
4.1.4 Nitrogen 46
4.1.5 Gas Proportion 47
4.2 Temperature 50
4.3 Summary 51
References 51

5 Atmospheric PECVD Growth of Vertically-Oriented Graphene 55
5.1 How Pressure Affects the Productivity of VG During PECVD 55
5.2 Atmospheric PECVD Growth of VG on Various Substrates 57
5.2.1 Pretreatment 58
5.2.2 Patterned Growth 58
5.2.3 Cylindrical Substrate 59
5.2.4 CNT Substrate 61
5.3 Summary 63
References 63

6 Vertically-Oriented Graphene for Sensing and Environmental Applications 67
6.1 VG-Based Biosensors 68
6.1.1 Electronic Biosensors 68
6.1.2 Electrochemical Biosensors 70
6.2 VG-Based Gas Sensors 72
6.3 VG-Based Corona Discharge 75
6.4 Summary 76
References 76

7 Vertically-Oriented Graphene for Supercapacitors 79
7.1 VG-Based Active Materials for EDLCs 81
7.2 VG-Based Active Materials for Pseudo-Capacitors 87
7.3 VGs for Bridging Active Materials and Current Collectors 92
7.4 Summary 94
References 94

8 Vertically-Oriented Graphene for Other Energy Storage and Conversion Applications 97
8.1 VG-Based Active Materials for Lithium-Ion Batteries 98
8.2 VG-Based Active Materials for Vanadium Redox Flow Batteries ... 101
8.3 VG-Based Active Materials for Fuel Cells 103
8.4 VG-Based Active Materials for Solar Cells 106
8.5 Summary 107
References 107

9 Conclusions and Outlook 109
9.1 Conclusions 110
9.2 Challenges and Outlook 110
9.3 Summary 112

Index 113

Abbreviations

1D	One-dimensional
2D	Two-dimensional
3D	Three-dimensional
AA	Ascorbic acid
AC	Activated carbon
a-C	Amorphous carbon
ACF	Activated carbon fiber
$Al(acac)_3$	Aluminum acetylacetonate
Au NP	Gold nanoparticle
CAG	Carbon aerogel
CB	Conduction band
CC	Carbon cloth
CCP	Capacitively coupled plasma
CE	Coulombic efficiency
CNT	Carbon nanotube
CNW	Carbon nanowall
CP	Carbon paper
CV	Cyclic voltammetry
CVD	Chemical vapor deposition
DA	Dopamine
dc	Direct current
DFT	Density functional theory
DOS	Density of states
dsDNA	Double-stranded DNA
DSSC	Dye-sensitized solar cell
EDL	Electric double layer
EDLCs	EDL capacitors
EEDF	Electron energy distribution function
ET	Electron transfer
FET	Field-effect transistor

FLG	Few layered graphene
FTO	Fluorine-doped tin oxide
FWHM	Full width at half maximum
HBGN	Highly branched graphene nanosheet
HER	Hydrogen evolution reaction
HRP	Horseradish peroxidase
HRTEM	High-resolution TEM
ICP	Inductively coupled plasma
IgG	Immunoglobulin G
IgM	Immunoglobulin M
LIB	Lithium-ion battery
MOR	Methanol oxidation reaction
MPECVD	Microwave plasma-enhanced chemical vapor deposition
MW	Microwave
MWCNT	Multiwalled CNT
OES	Optical emission spectroscopy
PANI	Polyaniline
PCE	Power conversion efficiency
PECVD	Plasma-enhanced chemical vapor deposition
POG	Perpendicularly oriented graphene
PV	Photovoltaic
PVA	Polyvinyl alcohol
RF	Radio frequency
RGO	Reduced graphene oxide
SEI	Solid electrolyte interface
SEM	Scanning electron microscope
SS	Stainless steel
T-CVD	Thermal chemical vapor deposition
TE	Transverse electric
TEM	Transmission electron microscope
TM	Transverse magnetic
UA	Uric acid
VACNF	Vertically aligned carbon nanofiber
VB	Valence band
VG	Vertically-oriented graphene
VHF	Very high-frequency
VLS	Vapor–liquid–solid
VRFB	Vanadium redox flow battery
VSS	Vapor–solid–solid

Chapter 1
Introduction

Abstract Graphene, a single layer of sp^2 hybridized carbon atoms, is a new 'star' in carbon allotropes. This one-atom-thick material has attracted enormous attention across the research community in the last decade due to its unique two-dimensional (2-D) structure and outstanding properties that are promising for both fundamental and applied research. Compared with conventional graphene sheets randomly laid down on a substrate, vertically-oriented graphene (VG) sheets possess advantageous characteristics, including exposed sharp edges, non-stacking morphology, and a large surface-to-volume ratio, and have thus shown great potential in various environmental and energy devices/systems. Plasma-enhanced chemical vapor deposition (PECVD) has been used widely as an effective method for VG synthesis. However, it remains a challenge for the controllable, large-scale, and low-cost growth of VG with desirable characteristics for specific applications. This chapter first provides a brief introduction to graphene, VG, and PECVD. Within that context, the main objective and the structure of the book are then presented.

Keywords Carbon · Graphene · Vertically-oriented graphene · Hybridization · Plasma-enhanced chemical vapor deposition · Spatial alignment · Electron mobility

1.1 Versatility of Carbon

Carbon is the sixth element in the periodic table and is a fascinating one in many ways. One interesting aspect of carbon lies in the flexibility of configuring electronic states of a carbon atom and thus bonding between neighboring carbon atoms. Carbon has a $1s^2 2s^2 2p^2$ ground-state electron configuration. The four valence orbitals ($2s$, $2p_x$, $2p_y$, and $2p_z$) in carbon have very similar energies and can readily mix with each other to form sp, sp^2, or sp^3 hybridizations. This unique ability to hybridize enables carbon to have different structural arrangements: sp, sp^2, and sp^3 hybridizations give rise to chain (one-dimensional/1-D), planar (2-D), and tetrahedral (three-dimensional/3-D) structures, respectively.

J. Chen et al., *Vertically-Oriented Graphene*, DOI 10.1007/978-3-319-15302-5_1

For example, for a carbon atom in graphene, the valence orbitals ($2s$, $2p_x$, $2p_y$, and $2p_z$) are sp^2 hybridized (i.e., the mixing of the single $2s$ orbital with two $2p$ orbitals), resulting in three planar σ orbitals separated by 120° with each other and one remaining $2p$ orbital ($2p_z$) oriented along the axis perpendicular to the graphene plane. Each of the three sp^2 orbitals of a carbon atom forms an σ bond with three neighboring carbon atoms, thereby forming the hexagonal structure of graphene.

1.2 Graphene: A Two-Dimensional Carbon Nanostructure

Graphene is a flat monolayer of sp^2-bonded carbon atoms that are tightly packed into a 2-D honeycomb lattice with a thickness of 0.335 nm. The last decade has witnessed explosive interest and growth in the research on graphene. As one example, the 2010 Nobel Prize in Physics was awarded to two scientists (Andre Geim and Konstantin Novoselov, http://www.nobelprize.org/nobel_prizes/physics/laureates/2010/) for their groundbreaking work on graphene, which further elevated graphene to the status of a 'star' material for the worldwide scientific community. The unprecedented enthusiasm about graphene is not surprising at all, given the exceptional mechanical [1], thermal [2], optical [3], and electrical [4–6] properties of graphene. For example, graphene was reported to outperform carbon nanotubes (CNTs) in room-temperature heat conduction [2]. Graphene has extremely high electron mobility at room temperature with resistivity of 10^{-6} Ω cm, which is comparable to, or even lower than that of, silver (1.62×10^{-6} Ω cm), a material widely known for its lowest resistivity at room temperature. Moreover, electron transport in graphene remains ballistic up to 0.3 μm at 300 K [6]. The extraordinary and outstanding properties of graphene make it a promising candidate for a wide range of applications, including field-effect transistors (FETs) [7, 8], composite materials [9, 10], field emitters [11], chemical sensors [12–17], biosensors [18–20], hydrogen storage media [21–23], and transparent conductive electrodes [24–28]. A comprehensive account of the recent history of graphene research can be found in a recent review paper by Ruoff and coworkers [29].

1.3 Vertically-Oriented Graphene and Its Growth

The importance of the spatial alignment of 1-D nanostructures, such as nanowires, nanorods, and nanotubes, to their applications has been well recognized [30–32]. A well-known example is the vertically-aligned CNT array [33, 34], which has been extensively demonstrated to show advantages over CNT powders and randomly oriented CNT mats in various applications such as field emitters, electromechanical actuators, gas sensors, and catalysis [35–39].

This 'spatial alignment effect' can be further extended to 2-D nanostructures, such as graphene, a lattice of sp^2 carbon atoms densely packed into a hexagonal structure and covalently bonded along two planar directions [4, 40]. For surface-bound single-layer graphene or stacks of graphene sheets, the orientation could be either horizontal or vertical, corresponding to graphene in parallel with or perpendicular to the synthesis/coating substrate, respectively. VG nanosheets, i.e., the so-called carbon/graphene nanowalls [41–63], carbon/graphene nanosheets [64–71], carbon/graphene nanoflakes [72, 73], and carbon nanoflowers [74], are a class of networks of graphitic platelets that are typically oriented vertically on a substrate. In these nanostructures, an individual VG nanosheet usually has lateral and vertical dimensions of 0.1 to tens of micrometers and a thickness of only a few nanometers (even less than 1 nm [75]). Each nanosheet consists of graphene with a layer number of 1–10 and an interlayer spacing of ~0.34 to ~0.37 nm [76]. Compared with conventional horizontally oriented graphene, the rising interest in the application of VG nanosheets initially stems from their unique orientation, exposed sharp edges, non-stacking morphology, and large surface-to-volume ratio. Until now, emerging applications of VG or its derivatives mainly included field emitters, atmospheric nanoscale corona discharges, gas sensors and biosensors, supercapacitors, lithium-ion batteries, fuel cells (catalyst supports), dye-sensitized solar cells, and electrochemical transducers.

It is noteworthy that these applications usually call for different morphologies and structures of VG nanosheets. For example, in the case of using VG as the electrode of atmospheric nanoscale corona discharges, exposed sharp edges and excellent electrical conductivity of graphene were key requirements for the electric field enhancement; meanwhile, a moderate intersheet/interlayer spacing was preferred for the simultaneous improvement in the discharge current and minimizing the electrostatic screening effect [77]. When VG was used as the active electrode material in electric double-layer capacitors (i.e., the so-called supercapacitors) or lithium-ion batteries, a huge specific surface area was desired for possible massive ion loading, and the intersheet/interlayer spacing needed to be well adjusted to optimize the ion diffusion, adsorption, or intercalation aiming at better energy storage performance. For the application of photovoltaics where the outstanding electrical conductivity of VG was used, a three-dimensional structure with highly branched nanosheet morphology could make it an excellent potential candidate for a light-scattering counter electrode [78]. Finally, the high surface area and abundant surface defects have been demonstrated to benefit gas-sensing applications [70]. Consequently, broad applications of VG call for controllable growth of VG with desirable characteristics, which will be the main topic of the current book.

With only a few exceptions that use glassy carbon or graphite sputtering [79] and substrate sputtering [80] techniques, chemical vapor deposition (CVD), especially the plasma-enhanced CVD (PECVD), has emerged as a key method for VG synthesis. Compared with thermal chemical vapor deposition (T-CVD) growth, PECVD offers several advantages, including a lower substrate temperature, higher growth selectivity, and better control in nanostructure ordering/patterning, due to the presence of energetic electrons, excited molecules and atoms, free radicals, photons, and other active species in the plasma region, which makes PECVD a popular method for VG growth. On the other hand, compared with

T-CVD methods primarily based on neutral gas chemistry, the growth of VG using PECVD is a more complex process, and the morphology and the structure of the as-produced VG sheets are strongly influenced by both the plasma source and a series of operating parameters.

1.4 Objectives and Organization of the Book

PECVD has emerged as a key method for VG synthesis; however, controllable growth of VG with desirable characteristics for specific applications remains a challenge. This book attempts to (i) summarize unique properties of VG; (ii) review the state-of-the-art research on PECVD growth of VG nanosheets; (iii) provide guidelines on the design of plasma sources and operating parameters; (iv) demonstrate the potential of VG for a wide range of applications; and (v) offer a perspective on outstanding challenges that need to be addressed prior to enabling commercial applications of VG.

This book is organized into nine chapters, and the topical organization is outlined in Fig. 1.1. After providing some background information in this chapter, the book gives an introduction of VG properties in Chap. 2, where the reader is made aware of some fundamental and unusual aspects of VG. We believe a good understanding of the VG's unique properties is critical to appreciating why it has attracted extensive attention and its potential in various applications. Chapters 3, 4, and 5 are intended to provide the reader with a comprehensive picture of the VG synthesis using PECVD. Chapter 3 discusses the VG growth mechanism during a typical PECVD process and summarizes three types of plasma sources (i.e., microwave, radio frequency, and direct current plasmas) generally used in existing PECVD systems. Chapter 4 covers the importance of two operating parameters (i.e., precursor and temperature) in the PECVD synthesis of VG. Chapter 5 is devoted to the atmospheric PECVD growth of VG on various substrates, which is deemed critical for the large-scale production of high-quality VG at a low cost. Chapters 6 through 8 offer an overview of VG applications for various environmental and energy devices/systems. Specifically, Chap. 6 is focused on VG's potential as a novel sensing element in biosensors and gas sensors and as a 'green' discharge electrode for corona discharge. In Chap. 7, we introduce and discuss the use of VG-based materials as electrodes in supercapacitors, which are energy storage devices featuring excellent charge/discharge rates, long cycle life, and environmental friendliness. Chapter 8 discusses the use of VG in some other electrochemical energy storage devices (e.g., lithium-ion batteries and vanadium redox flow batteries) and energy conversion devices (e.g., fuel cells and dye-sensitized solar cells). The book ends with a discussion on challenges and future directions for the further development of PECVD growth of VG in Chap. 9.

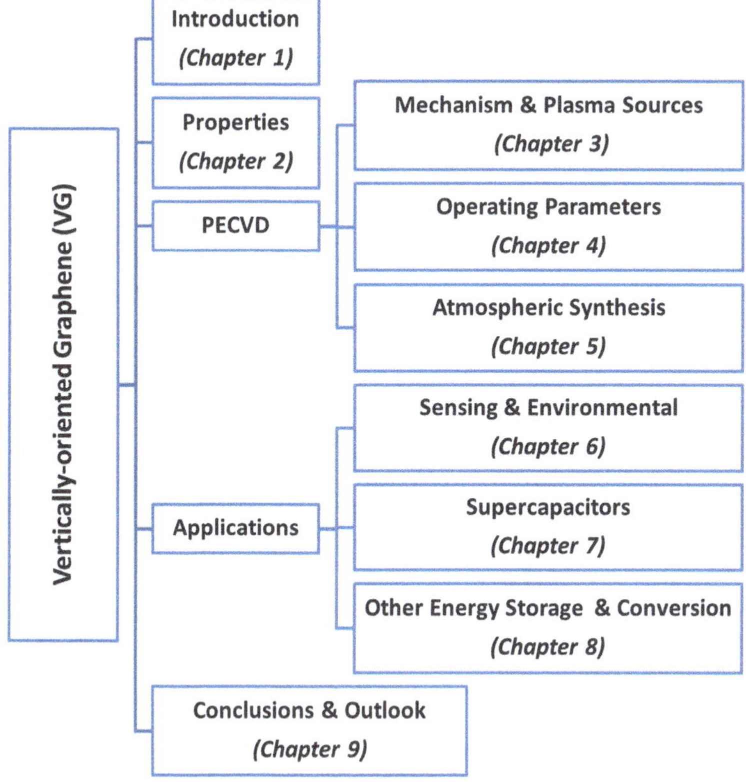

Fig. 1.1 The topical organization of the book

1.5 Summary

In this chapter, we have provided some background information about graphene, VG, and PECVD and also outlined the objective and main topics to be covered in this book. The versatility of carbon has resulted in many amazing structures, including a series of nanomaterials, e.g., fullerenes, CNTs, and graphene. VG has not only the general properties of intrinsic graphene, but also beneficial characteristics (e.g., sharp edges, non-stacking morphology, and high specific surface area) because of the aligned arrangement, thus holding great potential for various applications. Although PECVD has been an effective and widely used approach for VG synthesis, it remains challenging for the controllable, large-scale, and low-cost growth of VG with desirable characteristics for intended applications, which has been the motivation to prepare this book.

References

1. Frank, I. W., Tanenbaum, D. M., Van der Zande, A. M., & McEuen, P. L. (2007). Mechanical properties of suspended graphene sheets. *Journal of Vacuum Science and Technology B, 25*, 2558–2561.
2. Balandin, A. A., Ghosh, S., Bao, W. Z., Calizo, I., Teweldebrhan, D., Miao, F., et al. (2008). Superior thermal conductivity of single-layer graphene. *Nano Letters, 8*(3), 902–907.
3. Nair, R. R., Blake, P., Grigorenko, A. N., Novoselov, K. S., Booth, T. J., Stauber, T., et al. (2008). Fine structure constant defines visual transparency of graphene. *Science, 320*(5881), 1308–1308.
4. Novoselov, K. S., Geim, A. K., Morozov, S. V., Jiang, D., Zhang, Y., Dubonos, S. V., et al. (2004). Electric field effect in atomically thin carbon films. *Science, 306*(5696), 666–669.
5. Zhang, Y. B., Tan, Y. W., Stormer, H. L., & Kim, P. (2005). Experimental observation of the quantum Hall effect and Berry's phase in graphene. *Nature, 438*(7065), 201–204.
6. Geim, A. K., & Novoselov, K. S. (2007). The rise of graphene. *Nature Materials, 6*(3), 183–191.
7. Li, X. L., Wang, X. R., Zhang, L., Lee, S. W., & Dai, H. J. (2008). Chemically derived, ultrasmooth graphene nanoribbon semiconductors. *Science, 319*(5867), 1229–1232.
8. Semenov, Y. G., Kim, K. W. & Zavada, J. M. (2007). Spin field effect transistor with a graphene channel. *Applied Physics Letters, 91*(15), 153105.
9. Stankovich, S., Dikin, D. A., Dommett, G. H. B., Kohlhaas, K. M., Zimney, E. J., Stach, E. A., et al. (2006). Graphene-based composite materials. *Nature, 442*(7100), 282–286.
10. Ramanathan, T., Abdala, A. A., Stankovich, S., Dikin, D. A., Herrera-Alonso, M., Piner, R. D., et al. (2008). Functionalized graphene sheets for polymer nanocomposites. *Nature Nanotechnology, 3*(6), 327–331.
11. Wang, S. G., Wang, J. J., Miraldo, P., Zhu, M. Y., Outlaw, R., Hou, K., et al. (2006). High field emission reproducibility and stability of carbon nanosheets and nanosheet-based backgated triode emission devices. *Applied Physics Letters, 89*(18), 183103.
12. Schedin, F., Geim, A. K., Morozov, S. V., Hill, E. W., Blake, P., Katsnelson, M. I., et al. (2007). Detection of individual gas molecules adsorbed on graphene. *Nature Materials, 6*(9), 652–655.
13. Robinson, J. T., Perkins, F. K., Snow, E. S., Wei, Z. Q., & Sheehan, P. E. (2008). Reduced graphene oxide molecular sensors. *Nano Letters, 8*(10), 3137–3140.
14. Lu, G. H., Ocola, L. E., & Chen, J. H. (2009). Gas detection using low-temperature reduced graphene oxide sheets. *Applied Physics Letters, 94*(8), 083111.
15. Lu, G. H., Park, S., Yu, K. H., Ruoff, R. S., Ocola, L. E., Chen, J. H., et al. (2011). Toward practical gas sensing with highly reduced graphene oxide: A new signal processing method to circumvent run-to-run and device-to-device variations. *ACS Nano, 5*(2), 1154–1164.
16. Fowler, J. D., Allen, M. J., Tung, V. C., Yang, Y., Kaner, R. B., & Weiller, B. H. (2009). Practical chemical sensors from chemically derived graphene. *ACS Nano, 3*(2), 301–306.
17. Vedala, H., Sorescu, D. C., Kotchey, G. P., & Star, A. (2011). Chemical sensitivity of graphene edges decorated with metal nanoparticles. *Nano Letters, 11*(6), 2342–2347.
18. Mao, S., Lu, G. H., Yu, K. H., Bo, Z., & Chen, J. H. (2010). Specific protein detection using thermally reduced graphene oxide sheet decorated with gold nanoparticle-antibody conjugates. *Advanced Materials, 22*(32), 3521–3526.
19. Ohno, Y., Maehashi, K., & Matsumoto, K. (2010). Label-free biosensors based on aptamer-modified graphene field-effect transistors. *Journal of the American Chemical Society, 132*(51), 18012–18013.
20. Balapanuru, J., Yang, J. X., Xiao, S., Bao, Q. L., Jahan, M., Polavarapu, L., et al. (2010). A graphene oxide-organic dye ionic complex with DNA-sensing and optical-limiting properties. *Angewandte Chemie-International Edition, 49*(37), 6549–6553.
21. Patchkovskii, S., Tse, J. S., Yurchenko, S. N., Zhechkov, L., Heine, T., & Seifert, G. (2005). Graphene nanostructures as tunable storage media for molecular hydrogen. *Proceedings of the National Academy of Sciences of the United States of America, 102*(30), 10439–10444.

22. Park, N., Hong, S., Kim, G., & Jhi, S. H. (2007). Computational study of hydrogen storage characteristics of covalent-bonded graphenes. *Journal of the American Chemical Society, 129*(29), 8999–9003.
23. Boukhvalov, D. W., Katsnelson, M. I., & Lichtenstein, A. I. (2008). Hydrogen on graphene: Electronic structure, total energy, structural distortions and magnetism from first-principles calculations. *Physical Review B, 77*(3), 035427.
24. Watcharotone, S., Dikin, D. A., Stankovich, S., Piner, R., Jung, I., Dommett, G. H. B., et al. (2007). Graphene-silica composite thin films as transparent conductors. *Nano Letters, 7*(7), 1888–1892.
25. Wang, X., Zhi, L. J., & Mullen, K. (2008). Transparent, conductive graphene electrodes for dye-sensitized solar cells. *Nano Letters, 8*(1), 323–327.
26. Blake, P., Brimicombe, P. D., Nair, R. R., Booth, T. J., Jiang, D., Schedin, F., et al. (2008). Graphene-based liquid crystal device. *Nano Letters, 8*(6), 1704–1708.
27. Bae, S., Kim, H., Lee, Y., Xu, X. F., Park, J. S., Zheng, Y., et al. (2010). Roll-to-roll production of 30-inch graphene films for transparent electrodes. *Nature Nanotechnology, 5*(8), 574–578.
28. De, S., King, P. J., Lotya, M., O'Neill, A., Doherty, E. M., Hernandez, Y., et al. (2010). Flexible, transparent, conducting films of randomly stacked graphene from surfactant-stabilized, oxide-free graphene dispersions. *Small, 6*(3), 458–464.
29. Dreyer, D. R., Ruoff, R. S., & Bielawski, C. W. (2010). From conception to realization: An historial account of graphene and some perspectives for its future. *Angewandte Chemie-International Edition, 49*(49), 9336–9344.
30. Melechko, A. V., Merkulov, V. I., McKnight, T. E., Guillorn, M. A., Klein, K. L., Lowndes, D. H., et al. (2005). Vertically aligned carbon nanofibers and related structures: Controlled synthesis and directed assembly. *Journal of Applied Physics, 97*(4), 041301.
31. Wang, X. D., Summers, C. J., & Wang, Z. L. (2004). Large-scale hexagonal-patterned growth of aligned ZnO nanorods for nano-optoelectronics and nanosensor arrays. *Nano Letters, 4*(3), 423–426.
32. Whang, D., Jin, S., Wu, Y., & Lieber, C. M. (2003). Large-scale hierarchical organization of nanowire arrays for integrated nanosystems. *Nano Letters, 3*(9), 1255–1259.
33. Ajayan, P. M., Stephan, O., Colliex, C., & Trauth, D. (1994). Aligned carbon nanotube arrays formed by cutting a polymer resin-nanotube composite. *Science, 265*(5176), 1212–1214.
34. Li, W. Z., Xie, S. S., Qian, L. X., Chang, B. H., Zou, B. S., Zhou, W. Y., et al. (1996). Large-scale synthesis of aligned carbon nanotubes. *Science, 274*(5293), 1701–1703.
35. Fan, S. S., Chapline, M. G., Franklin, N. R., Tombler, T. W., Cassell, A. M., & Dai, H. J. (1999). Self-oriented regular arrays of carbon nanotubes and their field emission properties. *Science, 283*(5401), 512–514.
36. Baughman, R. H., Cui, C. X., Zakhidov, A. A., Iqbal, Z., Barisci, J. N., Spinks, G. M., et al. (1999). Carbon nanotube actuators. *Science, 284*(5418), 1340–1344.
37. Nguyen, C. V., Delzeit, L., Cassell, A. M., Li, J., Han, J., & Meyyappan, M. (2002). Preparation of nucleic acid functionalized carbon nanotube arrays. *Nano Letters, 2*(10), 1079–1081.
38. Janowska, I., Wine, G., Ledoux, M.-J., & Pham-Huu, C. (2007). Structured silica reactor with aligned carbon nanotubes as catalyst support for liquid-phase reaction. *Journal of Molecular Catalysis a-Chemical, 267*(1–2), 92–97.
39. Janowska, I., Hajiesmaili, S., Begin, D., Keller, V., Keller, N., Ledoux, M.-J., et al. (2009). Macronized aligned carbon nanotubes for use as catalyst support and ceramic nanoporous membrane template. *Catalysis Today, 145*(1–2), 76–84.
40. Novoselov, K. S., Geim, A. K., Morozov, S. V., Jiang, D., Katsnelson, M. I., Grigorieva, I. V., et al. (2005). Two-dimensional gas of massless Dirac fermions in graphene. *Nature, 438*(7065), 197–200.
41. Dikonimos, T., Giorgi, L., Giorgi, R., Lisi, N., Salernitano, E., & Rossi, R. (2007). DC plasma enhanced growth of oriented carbon nanowall films by HFCVD. *Diamond and Related Materials, 16*(4–7), 1240–1243.
42. Hiramatsu, M., Shiji, K., Amano, H., & Hori, M. (2004). Fabrication of vertically aligned carbon nanowalls using capacitively coupled plasma-enhanced chemical vapor deposition assisted by hydrogen radical injection. *Applied Physics Letters, 84*(23), 4708–4710.

43. Jain, H. G., Karacuban, H., Krix, D., Becker, H.-W., Nienhaus, H., & Buck, V. (2011). Carbon nanowalls deposited by inductively coupled plasma enhanced chemical vapor deposition using aluminum acetylacetonate as precursor. *Carbon, 49*(15), 4987–4995.
44. Luais, E., Boujtita, M., Gohier, A., Tailleur, A., Casimirius, S., Djouadi, M. A., et al. (2009). Carbon nanowalls as material for electrochemical transducers. *Applied Physics Letters, 95*(1), 014104.
45. Malesevic, A., Vizireanu, S., Kemps, R., Vanhulsel, A., Van Haesendonck, C., & Dinescu, G. (2007). Combined growth of carbon nanotubes and carbon nanowalls by plasma-enhanced chemical vapor deposition. *Carbon, 45*(15), 2932–2937.
46. Mori, T., Hiramatsu, M., Yamakawa, K., Takeda, K., & Hori, M. (2008). Fabrication of carbon nanowalls using electron beam excited plasma-enhanced chemical vapor deposition. *Diamond and Related Materials, 17*(7–10), 1513–1517.
47. Sato, G., Morio, T., Kato, T., & Hatakeyama, R. (2006). Fast growth of carbon nanowalls from pure methane using helicon plasma-enhanced chemical vapor deposition. *Japanese Journal of Applied Physics Part 1-Regular Papers Brief Communications & Review Papers, 45*(6A), 5210–5212.
48. Shimabukuro, S., Hatakeyama, Y., Takeuchi, M., Itoh, T., & Nonomura, S. (2008). Effect of hydrogen dilution in preparation of carbon nanowall by hot-wire CVD. *Thin Solid Films, 516*(5), 710–713.
49. Shin, S. C., Yoshimura, A., Matsuo, T., Mori, M., Tanimura, M., Ishihara, A., et al. (2011). Carbon nanowalls as platinum support for fuel cells. *Journal of Applied Physics, 110*(10), 104308.
50. Takeuchi, W., Ura, M., Hiramatsu, M., Tokuda, Y., Kano, H., & Hori, M. (2008). Electrical conduction control of carbon nanowalls. *Applied Physics Letters, 92*(21), 213103.
51. Wang, E. G., Guo, Z. G., Ma, J., Zhou, M. M., Pu, Y. K., Liu, S., et al. (2003). Optical emission spectroscopy study of the influence of nitrogen on carbon nanotube growth. *Carbon, 41*(9), 1827–1831.
52. Wu, Y. H., Yang, B. J., Zong, B. Y., Sun, H., Shen, Z. X., & Feng, Y. P. (2004). Carbon nanowalls and related materials. *Journal of Materials Chemistry, 14*(4), 469–477.
53. Yang, B. J., Wu, Y. H., Zong, B. Y., & Shen, Z. X. (2002). Electrochemical synthesis and characterization of magnetic nanoparticles on carbon nanowall templates. *Nano Letters, 2*(7), 751–754.
54. Chuang, A. T. H., Boskovic, B. O., & Robertson, J. (2006). Freestanding carbon nanowalls by microwave plasma-enhanced chemical vapour deposition. *Diamond and Related Materials, 15*(4–8), 1103–1106.
55. Eslami, P. A., Ghoranneviss, M., Moradi, S., Azar, P. A., Khorrami, S. A., & Laheghi, S. N. (2011). Growth of carbon nanowalls by thermal CVD on magnetron sputtered Fe thin film. *Fullerenes, Nanotubes, and Carbon Nanostructures, 19*(3), 237–249.
56. Kondo, S., Hori, M., Yamakawa, K., Den, S., Kano, H., & Hiramatsu, M. (2008). Highly reliable growth process of carbon nanowalls using radical injection plasma-enhanced chemical vapor deposition. *Journal of Vacuum Science and Technology B, 26*(4), 1294–1300.
57. Yu, K., Bo, Z., Lu, G., Mao, S., Cui, S., Zhu, Y., et al. (2011). Growth of carbon nanowalls at atmospheric pressure for one-step gas sensor fabrication. *Nanoscale Research Letters, 6*, 202.
58. Zhang, C., Hu, J., Wang, X., Zhang, X., Toyoda, H., Nagatsu, M., et al. (2012). High performance of carbon nanowall supported Pt catalyst for methanol electro-oxidation. *Carbon, 50*(10), 3731–3738.
59. Shimada, S., Teii, K., & Nakashima, M. (2010). Low threshold field emission from nitrogen-incorporated carbon nanowalls. *Diamond and Related Materials, 19*(7–9), 956–959.
60. Shemabukuro, S., Hatakeyama, Y., Takeuchi, M., Itoh, T., & Nonomura, S. (2008). Preparation of carbon nanowall by hot-wire chemical vapor deposition and effects of substrate heating temperature and filament temperature. *Japanese Journal of Applied Physics, 47*(11), 8635–8640.
61. Hojati-Talemi, P., & Simon, G. P. (2010). Preparation of graphene nanowalls by a simple microwave-based method. *Carbon, 48*(14), 3993–4000.
62. Teii, K., Shimada, S., Nakashima, M., & Chuang, A. T. H. (2009). Synthesis and electrical characterization of n-type carbon nanowalls. *Journal of Applied Physics, 106*(8), 084303.

63. Bo, Z., Yu, K., Lu, G., Wang, P., Mao, S., & Chen, J. (2011). Understanding growth of carbon nanowalls at atmospheric pressure using normal glow discharge plasma-enhanced chemical vapor deposition. *Carbon, 49*(6), 1849–1858.
64. Zhao, X., Tian, H., Zhu, M., Tian, K., Wang, J. J., Kang, F., et al. (2009). Carbon nanosheets as the electrode material in supercapacitors. *Journal of Power Sources, 194*(2), 1208–1212.
65. Wang, Z., Shoji, M., & Ogata, H. (2011). Carbon nanosheets by microwave plasma enhanced chemical vapor deposition in CH_4-Ar system. *Applied Surface Science, 257*(21), 9082–9085.
66. Tiwari, J. N., Tiwari, R. N., Singh, G., & Lin, K. L. (2011). Direct synthesis of vertically interconnected 3-D graphitic nanosheets on hemispherical carbon particles by microwave plasma CVD. *Plasmonics, 6*(1), 67–73.
67. Zhu, M. Y., Outlaw, R. A., Bagge-Hansen, M., Chen, H. J., & Manos, D. M. (2011). Enhanced field emission of vertically oriented carbon nanosheets synthesized by C_2H_2/H-2 plasma enhanced CVD. *Carbon, 49*(7), 2526–2531.
68. Chen, M. Y., Yeh, C. M., Syu, J. S., Hwang, J., & Kou, C. S. (2007). Field emission from carbon nanosheets on pyramidal Si(100). *Nanotechnology, 18*(18), 185706.
69. Wang, S., Wang, J., Miraldo, P., Zhu, M., Outlaw, R., Hou, K., et al. (2006). High field emission reproducibility and stability of carbon nanosheets and nanosheet-based backgated triode emission devices. *Applied Physics Letters, 89*(18), 183103.
70. Yu, K., Wang, P., Lu, G., Chen, K.-H., Bo, Z., & Chen, J. (2011). Patterning Vertically Oriented Graphene Sheets for Nanodevice Applications. *Journal of Physical Chemistry Letters, 2*(6), 537–542.
71. Kim, H., Wen, Z., Yu, K., Mao, O., & Chen, J. (2012). Straightforward fabrication of a highly branched graphene nanosheet array for a Li-ion battery anode. *Journal of Materials Chemistry, 22*(31), 15514–15518.
72. Shang, N. G., Papakonstantinou, P., McMullan, M., Chu, M., Stamboulis, A., Potenza, A., et al. (2008). Catalyst-Free Efficient Growth, Orientation and Biosensing Properties of Multilayer Graphene Nanoflake Films with Sharp Edge Planes. *Advanced Functional Materials, 18*(21), 3506–3514.
73. Shih, W.-C., Jeng, J.-M., Huang, C.-T., & Lo, J.-T. (2010). Fabrication of carbon nanoflakes by RF sputtering for field emission applications. *Vacuum, 84*(12), 1452–1456.
74. Hung, T.-C., Chen, C.-F., & Whang, W.-T. (2009). Deposition of Carbon Nanowall Flowers on Two-Dimensional Sheet for Electrochemical Capacitor Application. *Electrochemical and Solid State Letters, 12*(6), K41–K44.
75. Wang, J. J., Zhu, M. Y., Outlaw, R. A., Zhao, X., Manos, D. M., Holloway, B. C., et al. (2004). Free-standing subnanometer graphite sheets. *Applied Physics Letters, 85*(7), 1265–1267.
76. Bo, Z., Yang, Y., Chen, J., Yu, K., Yan, J., & Cen, K. (2013). Plasma-enhanced chemical vapor deposition synthesis of vertically oriented graphene nanosheets. *Nanoscale, 5*(12), 5180–5204.
77. Bo, Z., Yu, K., Lu, G., Cui, S., Mao, S., & Chen, J. (2011). Vertically oriented graphene sheets grown on metallic wires for greener corona discharges: lower power consumption and minimized ozone emission. *Energy & Environmental Science, 4*(7), 2525–2528.
78. Yu, K., Wen, Z., Pu, H., Lu, G., Bo, Z., Kim, H., et al. (2013). Hierarchical vertically oriented graphene as a catalytic counter electrode in dye-sensitized solar cells. *Journal of Materials Chemistry A, 1*(2), 188–193.
79. Deng, J.-H., Zheng, R.-T., Zhao, Y., & Cheng, G.-A. (2012). Vapor-Solid Growth of Few-Layer Graphene Using Radio Frequency Sputtering Deposition and Its Application on Field Emission. *ACS Nano, 6*(5), 3727–3733.
80. Yoo, J. J., Balakrishnan, K., Huang, J., Meunier, V., Sumpter, B. G., Srivastava, A., et al. (2011). Ultrathin Planar Graphene Supercapacitors. *Nano Letters, 11*(4), 1423–1427.

Chapter 2
The Properties of Vertically-Oriented Graphene

Abstract The unique properties of vertically-oriented graphene (VG) are discussed in this chapter. VG is intrinsically graphene, but it also possesses unique structural features, i.e., being arranged perpendicularly to the substrate surface. Therefore, VG possesses not only the properties of graphene but also some unique characteristics induced by its oriented arrangement. We start this chapter with a brief introduction of some general properties of graphene, which is deemed reasonable and necessary before we elaborate on the uniqueness of VG. To illustrate the attractive characteristics of VG, we compare VG with planar (or horizontal) graphene structures and emphasize the benefits that can be brought about due to VG's vertical orientation. The unique properties of VG are summarized at the end of this chapter. Understanding of the VG's uniqueness is critical to appreciating why VG has attracted so much interest and is also an essential step toward tailoring VG properties for various applications.

Keywords Ballistic electron mobility · Hybridization · Plasma-enhanced chemical vapor deposition · Orientation · Specific surface area · Thermal conductivity · Vertically-oriented graphene

2.1 General Characteristics of Graphene

The past decade has seen a fast-growing interest in graphene. This strong interest in graphene was initiated by some groundbreaking works in the early 2000s [1–3] and has been further enhanced by the Nobel Prize in Physics in 2010 [4]. However, graphene research started long before 2004 [5]. For example, the theoretical work in 1947 by Wallace [6] predicted that graphene (or single crystal graphite with interactions between planes being neglected, as the term "graphene" was unavailable at

Part of this chapter was adapted from our review article "Emerging Energy and Environmental Applications of Vertically-Oriented Graphenes," Chemical Society Reviews, 2015 (DOI: 10.1039/C4CS00352G)—Reproduced by permission of The Royal Society of Chemistry.

J. Chen et al., *Vertically-Oriented Graphene*, DOI 10.1007/978-3-319-15302-5_2

that time) might have extraordinary electronic characteristics (e.g., 100 times greater conductivity within a plane than between planes). In 1986, the term "graphene" was recommended to name a single carbon layer of the graphitic structure [5, 7, 8].

Graphene possesses many extraordinary properties and has been the subject of intense scientific interest. Exceptional values have been reported of: ballistic electron mobility (theoretical limit: ~2×10^5 $cm^2\ V^{-1}\ s^{-1}$ [9]; experimentally measured to be 2.53×10^5 $cm^2\ V^{-1}\ s^{-1}$ [10]), high thermal conductivity (5000 W/m-K) [11], Young's modulus (~1100 GPa), fracture strength (125 GPa) [12], a high specific surface area (~2630 m^2/g) [13], and optical absorption of exactly $\pi\alpha \approx 2.3\ \%$ (in the infrared limit, where α is the fine structure constant) [14]. To comprehend these outstanding properties, one must first look at the unusual electronic structure of graphene.

Electronic properties. As the sixth element in the periodic table, carbon has a $1s^2 2s^2 2p^2$ ground state electron configuration (i.e., two electrons in the inner shell and four in the outer shell). For a carbon atom in graphene, the four valence orbitals ($2s$, $2p_x$, $2p_y$, and $2p_z$) are sp^2 hybridized (i.e., the mixing of the single $2s$ orbital with two $2p$ orbitals), resulting in three planar σ orbitals separated by 120° with each other and one remaining $2p$ orbital ($2p_z$) oriented perpendicular to the graphene plane. Each of the three sp^2 orbitals of a carbon atom forms a σ bond (0.142 nm in length) with three neighboring carbon atoms on the 2D plane, forming the hexagonal structure of graphene with a basis of two atoms (A and B) per unit cell (Fig. 2.1a) that has the in-plane lattice constant of a = 0.246 nm and the thickness of 0.335 nm. The carbon-carbon σ bonds are strong covalent bonds, which are responsible for the lattice stability and for the elastic properties of graphene. The adjacent interaction among neighboring $2p_z$ orbitals develops into delocalized π (bonding) and π* (anti-bonding) bands, which form the valence band (VB) and the conduction band (CB), respectively (Fig. 2.1b).

The π and π* bands of graphene are degenerate at the corner (K point, or Dirac point) of the hexagonal Brillouin zone. For low energies associated with electron transport, the bands have a linear dispersion and the band structure can be regarded as two cones touching at E_D (the so-called Dirac crossing energy) (Fig. 2.1c). Since the VB and the CB touch at E_D, graphene has a zero bandgap

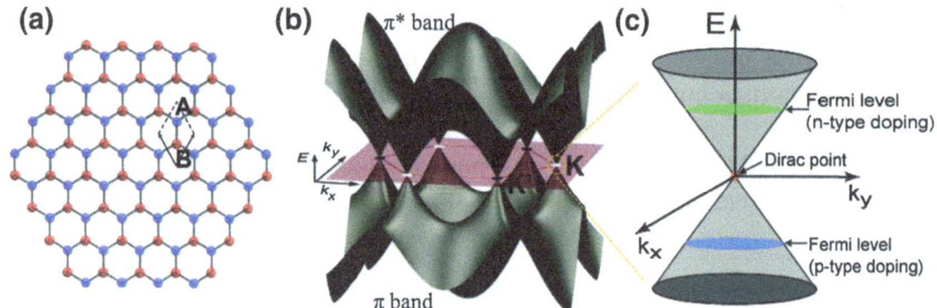

Fig. 2.1 **a** The hexagonal lattice of graphene has a basis of two carbon atoms (*A*, *B*) per unit cell. **b** The π and π* bands in graphene. **c** The linear dispersion and the band structure at the Dirac point. Reprinted with permission from [16]. Copyright 2010 American Chemical Society

and is thus typically labeled as a zero-gap semiconductor, or a semimetal. The extraordinary electronic properties of graphene are a direct result of the unusual band structure of graphene, i.e., a zero bandgap with linearly dispersing bands that touch at the Dirac point. The charge carrier mobility in graphene is very high, with experimental results of ~2.53×10^5 cm^2 V^{-1} s^{-1} [10] and theoretical prediction of ~2×10^5 cm^2 V^{-1} s^{-1} [9]. The charge carriers (electrons and holes) are able to travel a micrometer scale without scattering (known as ballistic transport) at room temperature [15]. Due to its atomic thickness, charge carrier transport through graphene is highly sensitive to adsorption/desorption of molecules, making graphene a promising material for both gas sensing and biosensing.

Mechanical properties. The covalent carbon-carbon bonds in graphene are very strong and nearly equivalent to the bonds holding carbon atoms together in diamond, giving graphene similar mechanical and thermal properties as diamond. Graphene is one of the strongest materials and has a Young's modulus of 1 TPa and an intrinsic strength of 130 GPa [12]. Since every atom in graphene is a surface atom, the specific surface area of graphene is extremely high and is theoretically predicted to be 2630 m^2 g^{-1}. Graphene is also very light, with a density of 0.77 mg/m^2; it would only take about 4 g of graphene to cover an American football field (110 m $\times$ 48.8 m).

2.2 Planar Graphene Versus Vertically-Oriented Graphene

The intrinsic properties of graphene can be unfavorably modulated when graphene sheets are laid down and in direct contact with a substrate. When graphene is laid on a substrate, the interaction between π electrons of graphene and the substrate electrons can considerably alter the electronic structure and lower the carrier mobility of graphene. For example, with SiO_2 as the substrate, the carrier mobility of graphene is limited to ~4×10^4 cm^2 V^{-1} s^{-1} at room temperature, which is much lower than the theoretical limit (~2×10^5 cm^2 V^{-1} s^{-1}). In addition, interactions with the underlying substrate are largely responsible for the presence of strong impurity scattering in graphene, which restricts the electron mean free path to less than a micron [17, 18].

When graphene sheets are placed randomly on a substrate, they tend to form irreversible agglomerates or restack due to the strong π–π stacking and van der Waals interactions, demolishing the potentials that could be offered by individual graphene sheets. Although a graphene sheet has exceptional electrical conductivity along its basal plane, the out-of-plane conductivity of graphene is much lower. Thus, numerous electrical contacts formed among the stacked graphene sheets contribute to additional, considerable electrical resistance. Moreover, when graphene sheets are stacked in a planar manner, direct access to the active surfaces is severely restricted.

Control over the orientation or arrangement of a nanostructure can provide advantages and additional leverage for certain applications. For example, oriented 1D nanomaterials, such as carbon nanotubes (CNTs), can outperform non-oriented counterparts in specific applications. It has been shown that vertically-oriented

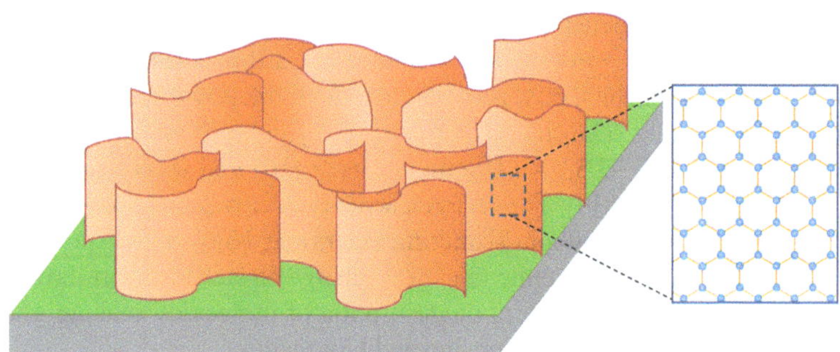

Fig. 2.2 Schematic of vertically-oriented graphene sheets

CNTs or CNT arrays have attractive characteristics as field emission electron sources for devices including flat panel displays, gas discharge tubes, and lamps [19]. When an electrical potential is applied between a CNT array and an anode, high local electric fields can be produced due to the very small radius of the CNT tip and the length of the CNT, which causes electrons to tunnel from the nanotube tip into the vacuum.

Similarly, when the orientation of graphene on a substrate is changed from being planar or horizontal (randomly oriented or parallel to the substrate surface) to being vertical (perpendicular to the substrate surface), i.e., to form VG (Fig. 2.2), it is promising to effectively harvest both intrinsic properties of graphene and additional characteristics due to the free standing arrangement.

2.3 Unique Properties of VG

In addition to general properties of graphene, VG sheets have some unique characteristics, making them significantly different in many aspects from the conventional horizontal, randomly oriented graphene sheets. And because of these unusual features, VG sheets possess a number of unique mechanical, chemical, electronic, electrochemical, and optoelectronic properties [20] that could benefit their potential use in a wide range of applications.

The first and probably most noticeable feature of VG sheets is the vertical orientation on the substrate, which improves the mechanical stability of graphene. While VG networks can have various morphologies, such as being petal-, turnstile-, maze-, and cauliflower-like [21–23], each VG nanosheet usually represents a free-standing, self-supported structure with rigid integrity by itself. This rigid structure preserves the mechanical stability of 2D graphene nanosheets, which would otherwise collapse and/or stack with each other in random directions, in part due to the strong van der Waals interactions. From the device perspective for electronic, optoelectronic, and electrochemical applications, the aligned structure of VG sheets can enable new designs and potentially improve the device performance. For example, the alignment of highly conductive graphene planes in parallel with the direction of charge transport in devices can result in a higher device efficiency. The vertical arrangement of VG sheets also facilitates the characterization of VG by using a scanning electron microscope (SEM), since their lateral dimensions are much larger than their thicknesses.

The second feature of VG sheets is a non-agglomerated morphology with a high specific surface area (or surface-to-volume ratio) and abundant open channels between the sheets. This feature can be taken as a direct result of VG's vertically-oriented structure, but it has significant implications worth being elaborated upon. Because VG sheets are non-agglomerated or non-stacked, the entire VG surface area can be readily accessed by gas/liquid molecules or ions. This enhanced accessibility is very critical to the performance of sensing and electrochemical devices that require maximized accessible surface area. One of the main reasons that graphene has attracted strong interest is its extremely high specific surface area due to the atomic thickness. However, re-arrangement (e.g., stacking) of horizontal graphene nanosheets can easily lead to a significant decrease in graphene's available surface area. The difficulty and challenge to preserve graphene's accessible surface area can be minimized to a large extent by using VG as an alternative. By carefully choosing growth parameters (e.g., plasma source, pressure, etc.) in a PECVD process, it is possible to adjust the inter-sheet spacing between the neighboring VG nanosheets from a few tens to several hundred nm and even larger [24–26]. Taking the advantage of this non-agglomerated structure, the specific surface area of the VG networks could reach a high value of ~1100 $m^2\ g^{-1}$ [27].

Third, VG sheets have long, exposed, ultra-thin, and reactive graphene edges, which are attractive for applications relying on the edge activity. An individual VG nanosheet typically has a tapered shape, i.e., its thickness reduces from a-few-graphene layers at the base to being atomically thin at the top [28]. The thin graphene layers in VG usually have an interlayer (002) spacing between 0.34 and 0.39 nm [29] and can be stacked in the Bernal AB configuration. However, rotating and disordered stacking orders are more often found in few-layer graphene sheets [28]. Recently, it was revealed that most of the VG edges are made of folded seamless graphene sheets and that only a relatively small fraction of the edges remain open during the plasma-based growth [30]. These active edges can boost the chemical and electrochemical activity of VG for sensing and electrochemical applications.

Fourth, VG sheets grown on a conductive surface minimize electrical resistances of the entire graphene network, as the extremely high in-plane conductivity of graphene can be effectively used, avoiding sheet-to-sheet resistance, and the contact resistance between VG and the substrate can be significantly reduced. In fact, first-principles calculations revealed that VG could possess electronic properties very similar to those of suspended graphene and ensure high carrier mobility as most of the π electrons of graphene are free of disturbance [31].

These unique morphological and structural characteristics make VG very attractive for many emerging energy and environmental applications in addition to the common field emission devices. For example, the large accessible surface area and high in-plane electrical conductivity can advance VG's use as a superior electrode material in various energy storage/conversion devices, such as supercapacitors, batteries, fuel cells, and dye-sensitized solar cells. The high density of open edges with controlled structural defects in VG can enhance the chemical and electrochemical activity and makes VG a promising sensing element for biosensors and gas sensors. The high aspect ratio and electrical conductivity of VG can facilitate the generation of atmospheric corona discharges with lower power consumption and reduced emission of hazardous ozone [32].

2.4 Summary

The properties of VG have been discussed in this chapter. Since VG is inherently graphene, general electronic and mechanical properties of graphene have been briefly introduced. On the other hand, being arranged perpendicularly to the surface of a substrate also induces a number of unique structural features for VG. These very unusual characteristics have been emphasized by comparing VG sheets with planar (or horizontal) graphene sheets. The unique properties of VG include: (i) improved mechanical stability owing to its vertical orientation; (ii) non-agglomerated morphology with a high specific surface area and abundant open channels; (iii) long, exposed, ultra-thin, and reactive graphene edges; and (iv) minimized contact resistance between VG and the substrate because of the direct growth of VG on a conductive surface. These features have made VG attractive for various applications.

References

1. Novoselov, K. S., Geim, A. K., Morozov, S. V., Jiang, D., Zhang, Y., Dubonos, S. V., et al. (2004). Electric field effect in atomically thin carbon films. *Science, 306*(5696), 666–669.
2. Novoselov, K. S., Geim, A. K., Morozov, S. V., Jiang, D., Katsnelson, M. I., Grigorieva, I. V., et al. (2005). Two-dimensional gas of massless Dirac fermions in graphene. *Nature, 438*(7065), 197–200.
3. Zhang, Y. B., Tan, Y. W., Stormer, H. L., & Kim, P. (2005). Experimental observation of the quantum hall effect and Berry's phase in graphene. *Nature, 438*(7065), 201–204.
4. Nobelprize.org. *The Nobel Prize in Physics 2010*. Nobel Media AB 2014; Available from: http://www.nobelprize.org/nobel_prizes/physics/laureates/2010/.
5. Dreyer, D. R., Ruoff, R. S., & Bielawski, C. W. (2010). From conception to realization: An historial account of graphene and some perspectives for its future. *Angewandte Chemie-International Edition, 49*(49), 9336–9344.
6. Wallace, P. R. (1947). The band theory of graphite. *Physical Review, 71*(9), 622–634.
7. Boehm, H. P., Setton, R., & Stumpp, E. (1986). Nomenclature and terminology of graphite-intercalation compounds. *Carbon, 24*(2), 241–245.
8. Boehm, H. P., Setton, R., & Stumpp, E. (1994). Nomenclature and terminology of graphite-intercalation compounds (Iupac recommendations 1994). *Pure and Applied Chemistry, 66*(9), 1893–1901.

9. Morozov, S. V., Novoselov, K. S., Katsnelson, M. I., Schedin, F., Elias, D. C., Jaszczak, J. A., & Geim, A. K. (2008). Giant intrinsic carrier mobilities in graphene and its bilayer. *Physical Review Letters, 100*(1), 016602.
10. Chen, J. H., Jang, C., Xiao, S. D., Ishigami, M., & Fuhrer, M. S. (2008). Intrinsic and extrinsic performance limits of graphene devices on SiO_2. *Nature Nanotechnology, 3*(4), 206–209.
11. Balandin, A. A., Ghosh, S., Bao, W. Z., Calizo, I., Teweldebrhan, D., Miao, F., & Lau, C. N. (2008). Superior thermal conductivity of single-layer graphene. *Nano Letters, 8*(3), 902–907.
12. Lee, C., Wei, X. D., Kysar, J. W., & Hone, J. (2008). Measurement of the elastic properties and intrinsic strength of monolayer graphene. *Science, 321*(5887), 385–388.
13. Stoller, M. D., Park, S. J., Zhu, Y. W., An, J. H., & Ruoff, R. S. (2008). Graphene-based ultracapacitors. *Nano Letters, 8*(10), 3498–3502.
14. Nair, R. R., Blake, P., Grigorenko, A. N., Novoselov, K. S., Booth, T. J., Stauber, T., et al. (2008). Fine structure constant defines visual transparency of graphene. *Science, 320*(5881), 1308–1308.
15. Mayorov, A. S., Gorbachev, R. V., Morozov, S. V., Britnell, L., Jalil, R., Ponomarenko, L. A., et al. (2011). Micrometer-scale ballistic transport in encapsulated graphene at room temperature. *Nano Letters, 11*(6), 2396–2399.
16. Avouris, P. (2010). Graphene: Electronic and photonic properties and devices. *Nano Letters, 10*(11), 4285–4294.
17. Bolotin, K. I., Sikes, K. J., Jiang, Z., Klima, M., Fudenberg, G., Hone, J., et al. (2008). Ultrahigh electron mobility in suspended graphene. *Solid State Communications, 146*(9–10), 351–355.
18. Hwang, E. H., Adam, S., & Sarma, S. D. (2007). Carrier transport in two-dimensional graphene layers. *Physical Review Letters, 98*(18), 186806.
19. Baughman, R. H., Zakhidov, A. A., & de Heer, W. A. (2002). Carbon nanotubes–the route toward applications. *Science, 297*(5582), 787–792.
20. Bo, Z., Mao, S., Han, Z. J., Cen, K., Chen, J., & Ostrikov, K. (2015). Emerging energy and environmental applications of vertically-oriented graphenes. *Chemical Society Reviews*, doi: 10.1039/C4CS00352G.
21. Wu, Y. H., Qiao, P. W., Chong, T. C., & Shen, Z. X. (2002). Carbon nanowalls grown by microwave plasma enhanced chemical vapor deposition. *Advanced Materials, 14*(1), 64–67.
22. Seo, D. H., Kumar, S., & Ostrikov, K. (2011). Control of morphology and electrical properties of self-organized graphenes in a plasma. *Carbon, 49*(13), 4331–4339.
23. Bo, Z., Zhu, W. G., Ma, W., Wen, Z. H., Shuai, X. R., Chen, J. H., et al. (2013). Vertically oriented graphene bridging active-layer/current-collector interface for ultrahigh rate supercapacitors. *Advan-ced Materials, 25*(40), 5799–5806.
24. Wang, J. J., Zhu, M. Y., Outlaw, R. A., Zhao, X., Manos, D. M., Holloway, B. C., & Mammana, V. P. (2004). Free-standing subnanometer graphite sheets. *Applied Physics Letters, 85*(7), 1265–1267.
25. Hiramatsu, M., Shiji, K., Amano, H., & Hori, M. (2004). Fabrication of vertically aligned carbon nanowalls using capacitively coupled plasma-enhanced chemical vapor deposition assisted by hydrogen radical injection. *Applied Physics Letters, 84*(23), 4708–4710.
26. Bo, Z., Yu, K., Lu, G., Wang, P., Mao, S., & Chen, J. (2011). Understanding growth of carbon nanowalls at atmospheric pressure using normal glow discharge plasma-enhanced chemical vapor deposition. *Carbon, 49*(6), 1849–1858.
27. Miller, J. R., Outlaw, R. A., & Holloway, B. C. (2010). Graphene double-layer capacitor with AC line-filtering performance. *Science, 329*(5999), 1637–1639.
28. Davami, K., Shaygan, M., Kheirabi, N., Zhao, J., Kovalenko, D. A., Rummeli, M. H., et al. (2014). Synthesis and characterization of carbon nanowalls on different substrates by radio frequency plasma enhanced chemical vapor deposition. *Carbon, 72*, 372–380.
29. Bo, Z., Yang, Y., Chen, J., Yu, K., Yan, J., & Cen, K. (2013). Plasma-enhanced chemical vapor deposition synthesis of vertically oriented graphene nanosheets. *Nanoscale, 5*(12), 5180–5204.

30. Zhao, J., Shaygan, M., Eckert, J., Meyyappan, M., & Rummeli, M. H. (2014). A growth mechanism for free-standing vertical graphene. *Nano Letters, 14*(6), 3064–3071.
31. Yuan, Q. H., Hu, H., Gao, J. F., Ding, F., Liu, Z. F., & Yakobson, B. I. (2011). Upright standing graphene formation on substrates. *Journal of the American Chemical Society, 133*(40), 16072–16079.
32. Bo, Z., Yu, K., Lu, G., Cui, S., Mao, S., & Chen, J. (2011). Vertically oriented graphene sheets grown on metallic wires for greener corona discharges: Lower power consumption and minimized ozone emission. *Energy & Environmental Science, 4*(7), 2525–2528.

Chapter 3
PECVD Synthesis of Vertically-Oriented Graphene: Mechanism and Plasma Sources

Abstract The plasma-enhanced chemical vapor deposition (PECVD) method is a key method for synthesizing vertically-oriented graphene (VG). Because the plasma region provides active species (e.g., energetic electrons, excited molecules and atoms, free radicals, and photons), PECVD offers several advantages in nanostructure synthesis, e.g., a relatively low substrate temperature, a high growth selectivity, and good control in nanostructure ordering/patterning. These features make PECVD the most suitable method for VG growth. On the other hand, the growth of VG using PECVD is a quite complex process due to the complexity of plasma chemistry. The morphology and structure of the VG sheets produced by PECVD are strongly dependent on the types of plasma sources and a series of operating parameters, such as feedstock gas type and composition, the substrate temperature, and the operating pressure. In this chapter, we first discuss the growth mechanism of VG in a PECVD process and then discuss how plasma sources affect the VG growth. Characterization of PECVD-produced VG from various plasma sources using Raman spectroscopy, a powerful tool to study carbon nanostructures, is also discussed in this chapter.

Keywords Electric field · Internal stress · Anisotropic growth effects · Microwave plasma · Radio frequency plasma · Direct current discharge · Raman spectroscopy

---

Part of this chapter was adapted from our review articles: "Plasma-Enhanced Chemical Vapor Deposition Synthesis of Vertically-oriented Graphene Nanosheets," *Nanoscale* **5**(12), 5180-5204, 2013 (DOI: 10.1039/C3NR33449J); and "Emerging Energy and Environmental Applications of Vertically-Oriented Graphenes," *Chemical Society Reviews*, 2015 (DOI: 10.1039/C4CS00352G)—Reproduced by permission of The Royal Society of Chemistry.

J. Chen et al., *Vertically-Oriented Graphene*, DOI 10.1007/978-3-319-15302-5_3

3.1 VG Growth Mechanism by PECVD

The growth mechanism of vertically-oriented graphene (VG) during the plasma-enhanced chemical vapor deposition (PECVD) process has been vigorously explored in recent years. However, it still remains elusive and even controversial in explaining some observations in the growth process. First, the complex nature of the PECVD combined with different VG synthesis conditions makes it very challenging toward a clear and conclusive understanding of VG growth. Many factors in the PECVD determine the VG growth and the final product VG structure. The substrate and precursor in the PECVD are believed to be the two dominant factors in the VG synthesis. In addition, several other factors such as plasma source and power, etching rate, surface temperature, and plasma pre-treatment can also affect the final VG structure [1]. The unique synthesis conditions and the distinctive morphological properties of VG also hinder the direct use of theories that have been successfully applied in describing the growth of other structurally-aligned nanomaterials. For example, the vapor–liquid–solid (VLS) or vapor–solid–solid (VSS) mechanisms have been widely used to elucidate the growth of vertically-oriented 1D nanotubes or nanowires from patterned catalysts, but they cannot be directly used to understand PECVD synthesis of VG as the VG growth requires no intentional catalyst. The nucleation mechanism for 2D thin film deposition is also of limited relevance because it describes continuous layers rather than networks of vertically-oriented, separated, wall-like structures such as VG nanosheets.

Although challenges remain, some progress has been made toward a clearer picture of VG growth in PECVD. Recent advances in time-resolved growth and microanalysis techniques now allow in-depth understanding of PECVD growth of VG, and several growth mechanisms have been proposed to advance the understanding of VG growth through PECVD methods. In general, the VG growth process in PECVD is believed to be involved with three critical, sequential steps: nucleation, growth, and termination [2]. These three steps can be described briefly as follows: (a) nucleation: the irregular cracks and dangling bonds on the substrate surface serve as nucleation sites, and a buffer layer is formed for the following VG growth; (b) growth: graphene nanosheets grow vertically under the influence of stress and/or a localized electric field, and carbon atoms are continuously incorporated into the open edges of a vertical graphene sheet; and (c) termination: the growth of VG finally stops upon the closure of the open edges, which is determined by the competition of material deposition and etching effects in the plasma [3].

The buffer layer, which is formed in the nucleation step, is usually made of either amorphous carbon (a-C) or carbide [4, 5]. Amorphous carbon is formed due to the large mismatch between the lattice parameters of the substrate material and the graphite, while a carbide layer is formed when the substrate can react with or dissolve carbon atoms [3]. A planar or onion-like graphitic layer is found between the amorphous carbon buffer layer and the vertical graphene nanosheets, as shown in Fig. 3.1a [6]. Once the buffer layer is formed and the VG nanosheets start to grow, VG sheets no longer show any substrate-dependent features, which explains why similar morphology has been observed for VG sheets grown on different types

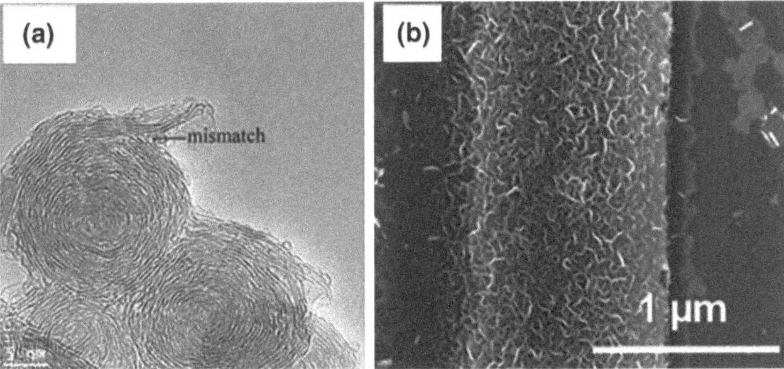

Fig. 3.1 VG growth mechanism: **a** transmission electron microscopy (TEM) image of a carbon onion with mismatched graphitic layers at the surface, which may initialize the VG growth. Reprinted with permission from [6]. Copyright 2014 American Chemical Society. **b** SEM image of VG sheets grown on an Au stripe due to the electric field effect. Reprinted with permission from [9]. Copyright 2011 American Chemical Society

of substrates. The amorphous carbon buffer layer can be etched away with etchant radicals such as H or OH groups, which benefits the growth of VG nanosheets [7].

With the formation of the buffer layer and nucleation sites, the graphene nanosheet starts to grow. The next essential question is why VG can grow vertically, instead of growing thickness-wise to form thicker graphene films, as often observed in other carbon-based nanostructures (e.g., multiwalled CNTs or MWCNTs). Based on experimental observations and theoretical investigations, the vertical growth is likely due to three main reasons: the electric field, the internal stress, and the anisotropic growth effects.

Electric field. The electric field in the plasma sheath can direct the growth of various oriented nanostructures (e.g., vertically standing CNTs or a CNT array) in the vicinity of the substrate surface [8]. Hence, the VG growth direction and spatial distribution are affected by the electric field in the plasma sheath. We have investigated the effect of the electric field on VG growth and found that modulating the local electric field above the substrate can effectively control the density and orientation of the VG networks. In the case of grounded conductive substrates, the electric field is normal to the substrate surface and is stronger near the edges and sharp points than near the flat surface. As shown in Fig. 3.1b, VG sheets were produced with a high density on the Au stripe (conductive), while neither VG nor amorphous carbon was found on the neighboring nonconductive SiO_2 surface [9]. This phenomenon can be explained by the fact that the electric field above the Au stripe, especially near the edges, is much stronger than that above the SiO_2 substrate. Customizing the surface electric field distribution thus opens an avenue to growing patterned VG structures for device applications. On the other hand, when the substrate is nonconductive or floating (disconnected from an external electric circuit) in the plasma, the relatively low electric field leads to much more irregular and random VG networks [5].

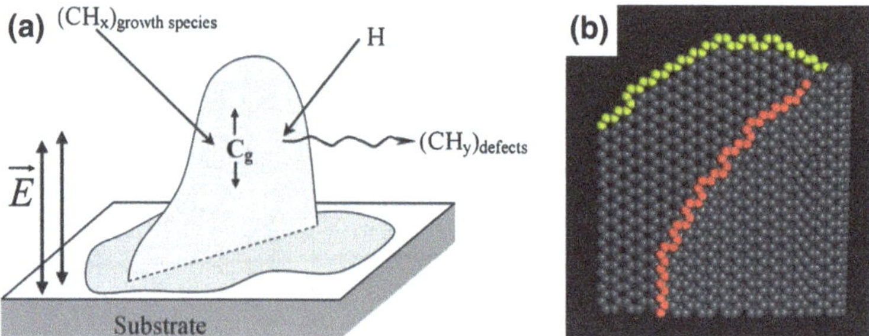

Fig. 3.2 **a** A schematic of VG growth controlled by the electric field and carbon surface diffusion. Reprinted with permission from [11]. Copyright 2007 Elsevier. **b** Atomistic model of a curved vertical graphene with active, growing edges (*highlighted in color*). Reprinted with permission from [6]. Copyright 2014 American Chemical Society

Internal stress. Internal stress arising from the temperature gradients, ion bombardment, and lattice mismatch between the substrate material and the graphitic material may cause defects or buckles in the buffer layer, which work as the nucleation sites for the VG growth. The initial growth of planar or onion-like graphitic layers eventually switches to upward growth of impinging graphene sheets, which releases the stress accumulated during the initial growth stage. The dissociated carbon species in the plasma will continuously provide radicals, ions, and neutrals to the open sites of vertically growing hexagonal lattices of VG sheets [5].

Anisotropic growth effects. The directional growth of VG sheets could also be attributed to the anisotropic growth effects. It has been proposed that the growth rates in the parallel and perpendicular directions to the graphene layer were different [10]. Specifically, the VG sheets oriented normally to the substrate usually grow faster than their randomly oriented counterparts, which is partly due to the surface diffusion of carbon atoms (Fig. 3.2a) [11]. Carbon-containing species, once landing on the surface of a growing nanosheet, will rapidly move along the sheet surface, reach the upper edge, and covalently bond to the edge atoms. On the contrary, the carbon-containing species diffusing to the substrate surface can be re-evaporated and desorbed from the surface because of the weak adsorption to the substrate. The high surface diffusion rates, caused by the large difference between the surface adsorption energy (~0.13 eV) and the surface diffusion energy (~1.7 eV) for a carbon atom (from the growth species) on the graphene surface, could assist carbon atoms to migrate along the graphene surface, leading to a vertical growth. In addition, due to the sharp features of the VG edge (stronger localized electric fields), carbon atoms also preferentially grow from the edges of VG [8]. As a result, the growth rate in the vertical direction is higher than that in the lateral direction.

Recently, a kinetic model supported by experimental investigations suggested that the VG growth can be considered as a step flow process where the nucleation takes place at the bottom [4, 6]. According to this model, the VG nucleation is triggered by the mismatch of graphitic carbon layers at either the buffer layer or the carbon onions that form on the surface. The growth of individual nanosheets is then determined by the number of layers nucleated from the bottom and the diffusion

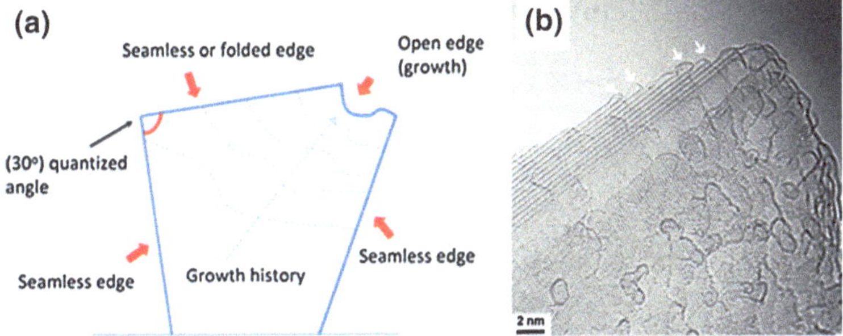

Fig. 3.3 **a** Schematic of VG with folded/seamless and open edges. Reprinted with permission from [4]. Copyright 2014 Elsevier. **b** TEM image of a VG sheet with the tapered shape and folded edges shown by the *arrows*. Reprinted with permission from [6]. Copyright 2014 American Chemical Society

rate of carbon atoms to each layer (Fig. 3.2b). This model also suggests that VG growth only occurs at open edges but not at folded or seamless edges [2], as shown in Fig. 3.3a. As the neighboring layers can form a closure and cease the growth, tapered VG nanosheets may form, as evidenced in the TEM image in Fig. 3.3b.

Although the PECVD growth mechanism of VG with gaseous precursors has received much attention, the VG growth from liquid or solid precursors is much less explored and currently lacks clear understanding. However, there are some apparent similarities in the growth kinetics when VG sheets are produced from liquid, solid, or gaseous precursors. The plasma first acts on the solid or liquid precursors by dehydrating them as a result of the plasma-related heating. The plasma then converts or decomposes the dehydrated precursors into smaller, more common carbon-containing species, regardless of the initial precursor, through interactions with plasma-generated ions and radicals. These species then act as the basic building units for the VG growth [12]. Of course, several points need to be investigated to fully understand the VG grown from different precursors. For example, why VG sheets grown from different precursors exhibit different adhesions to the substrate; and what exact surface reconstructions under the plasma exposure lead to the preferential growth of VG in the vertical direction. Future studies on the VG growth mechanism with solid or liquid precursors are needed and will bring more and deeper understanding on this interesting VG growth process with PECVD.

3.2 Plasma Sources

PECVD growth of VG can be accomplished using a variety of plasma sources, such as microwave (MW), radio frequency (RF), and direct current (dc) discharges with different reactor configurations. A few examples of PECVD systems are schematically shown in Fig. 3.4, and the typical operational parameters of the PECVD systems are summarized in Table 3.1.

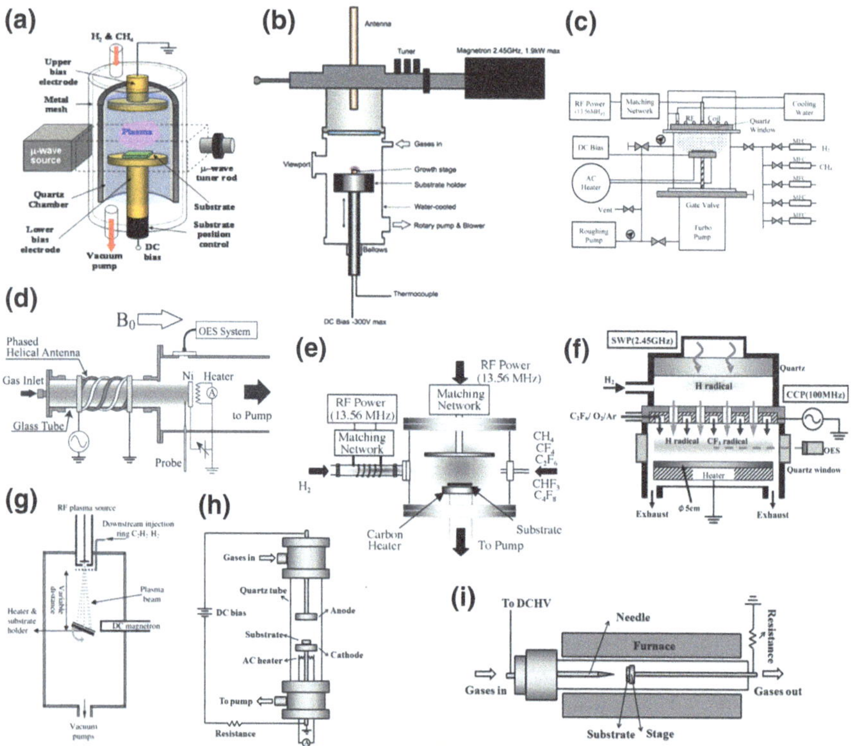

Fig. 3.4 Schematics of various PECVD systems for VG growth ([1]—Reproduced by permission of The Royal Society of Chemistry): **a** transverse electric (TE)-MW (Reprinted with permission from [48]. Copyright 2010 AIP Publishing LLC); **b** transverse magnetic (TM)-MW (Reprinted with permission from [47]. Copyright 2006 Elsevier); **c** inductively coupled plasma (ICP) (Reprinted with permission from [20]. Copyright 2004 Elsevier); **d** helicon plasma (Reprinted with permission from [21]. Copyright 2006 Japan Society of Applied Physics); **e** capacitively coupled plasma (CCP)+ICP (Reprinted with permission from [28]. Copyright 2005 Elsevier); **f** very-high-frequency (VHF) CCP+MW (Reprinted with permission from [29]. Copyright 2008 AIP Publishing LLC); **g** expanding CCP (Reprinted with permission from [26]. Copyright 2010 IOP Publishing); **h** parallel-plate dc glow discharge plasma; and **i** pin-to-plate normal glow discharge plasma

Surface-bound VG nanosheets were initially discovered during the fabrication of CNTs in 1997 by dc arc discharge evaporation of graphite in the presence of rarefied hydrogen gas [13]. Since then, a wide range of plasma sources with different power frequencies, such as MW plasma (f = 0.5–10 GHz; commonly 2.45 GHz), RF plasma (f = 1–500 MHz; commonly 13.56 MHz), dc plasma, and their combinations, have been developed for the VG synthesis. Besides gas discharge plasmas, the electron beam excited plasma has also been used for VG growth but with limited applications.

Table 3.1 Overview of PECVD processes for VG synthesis [1]—Reproduced by permission of The Royal Society of Chemistry

Plasma source	Precursor	Growth pressure (Pa)	Substrate temperature (°C)	Flow rate (sccm[a])	Gas ratio	References
Helicon	CH_4	0.04–2	700	–	–	[21]
ICP	CH_4	12	630–830	10	–	[20]
ICP	CH_4/Ar	0.3	400	30.4	16.4:14	[10]
TM-MW	CH_4/Ar	17.33	450–500	–	1:8	[49]
TM-MW	CH_4/N_2	5.32×10^3	1250	–	–	[39]
TM-MW	CH_4/N_2	5.32×10^3	>1000	–	–	[50]
TM-MW	$C_2H_2/N_2/Ar$	1.33×10^4	650–1050	200	0.5 % C_2H_2	[40]
TM-MW	$CH_4/N_2/Ar$	1.33×10^4	650–1050	200	4 % CH_4	[40]
ICP	CH_4/H_2	12	630–830	10	>1:9	[20]
ICP	CH_4/H_2	2.66–53.2	600–950	–	>1:19	[11]
ICP	CH_4/H_2	13.33	700	10	2:3	[41]
ICP	C_2H_2/H_2	4–5.33	550–600	5	4:1	[41]
TE-MW	CH_4/H_2	133	650–700	50	1:4	[16]
TE-MW	CH_4/H_2	220	550	–	1:20	[17]
TE-MW	CO/H_2	250	700	50	23:2	[51]
TM-MW	CH_4/H_2	5.32×10^3	700	200	1:8	[3]
TM-MW	$CH_4/H_2/Ar$	1.33×10^4	650	44	1:1:20	[52]
TM-MW	C_2H_2/NH_3	1.33×10^3	–	–	>1:1	[47]
TE-MW[b]	CH_4/CO_2	–	900	–	53:47	[53]
Expending CCP[b]	$C_2H_2/H_2/Ar$	–	700	1076	1:25:1050	[26]
CCP+ICP	CH_4/H_2	13.3	500	45	1:2	[28]
CCP+ICP	CF_4/H_2	13.3	500	45	1:2	[28]
CCP+ICP	CHF_3/H_2	13.3	500	45	1:2	[28]
CCP+ICP	C_2F_6/H_2	13.3	500	45	1:2	[28]
VHFCCP+MW	C_2F_6/H_2	13.3–1596	600	150	1:2	[29]
VHFCCP+MW	$C_2F_6/H_2/N_2$	13.3–1596	600	155	10:20:1	[30]
VHFCCP+MW	$C_2F_6/H_2/O_2$	13.3–1596	600	155	10:20:1	[52]
DC glow	CH_4/H_2	1×10^4	1000	–	1:9	[54]
DC glow	CH_4/H_2	9975	1000	–	8:92	[55]
DC glow[c]	CH_4/H_2	2.66×10^4	900–1000	50	3–8 % CH_4	[56]
DC glow[b]	$CH_4/H_2/Ar$	1.3	550–800	87	1:1.25:5	[33]
DC glow	$CH_4/H_2O/Ar$	1.01×10^5	700	1500	10 % CH_4[d]	[7]

[a] *sccm* standard cubic centimetre per minute
[b] Catalyst was used
[c] Substrate with MW/RF CVD treatment
[d] Relative humidity: ~40 %

3.2.1 Microwave Plasmas

MW plasma is a type of electrodeless gas discharge plasma with high frequency electromagnetic radiation in the GHz range. The wavelength of MW is in the centimeter range, which is comparable with the discharge system, so the interaction between the electromagnetic field and the plasma in MW discharges is quasi-optical. For MW confined by a reflective boundary (i.e., the so-called waveguide), there are two typical wave propagation modes, i.e., transverse electric (TE) mode and transverse magnetic (TM) mode, referred to as the electric field and magnetic field perpendicular to the direction of wave travel, respectively.

TE mode MW reactors driven by a rectangular waveguide have been extensively used for the VG synthesis. As shown in Fig. 3.4a, a 2.45 GHz MW reactor is coupled to the VG synthesis reactor vessel (a cylindrical quartz tube) via a traverse rectangular cavity waveguide. The incident electromagnetic wave formed in the waveguide interacts with the plasma generated in the discharge. This interaction results in partial dissipation, partial transmission, and reflection of the electromagnetic wave. To increase the effectiveness of electromagnetic wave coupling with the plasma column, the transmitted wave can be reflected back, which leads to the formation of a standing wave. A tuner is used to adjust the waveguide length, so that the standing wave electric field in the growth region is the strongest. Such special coupling techniques can increase the fraction of the electromagnetic energy absorbed in the plasma, which is important for practical applications of the discharge system. The TE-MW reactor can synthesize VG with a relatively facile setup; however, it also has disadvantages such as limited substrate temperature and the possible introduction of contamination from the outside container, since the wave is directly coupled to the quartz tube with a surface wave plasma mode. Furthermore, the operating power for VG synthesis using TE-MW reactors was limited to typically 60–500 W, since the quartz tube and the vacuum system could be damaged under a high operating temperature or a high MW power. In addition, the MW power cannot be easily confined, and the spread of MW can lead to a decreased growth efficiency and less uniformity of the morphology and the structure of as-grown VG.

A possible route to addressing the above problems with the TE-MW reactor is to use the TM-MW reactor, in which the dominant wave is converted from the TE mode in a rectangular waveguide to the TM mode in a cylindrical waveguide. As shown in Fig. 3.4b, an antenna is introduced at the top of the VG synthesis reactor vessel, a cylindrical cavity, through a coaxial port, to produce a more intense electric field on the central part of the substrate where a plasma ball is clearly observed. Different from the TE-MW reactor in which the substrate is immersed into the plasma region, the substrate of a TM-MW reactor is below the plasma ball and thus enables a controllable substrate temperature. A dielectric window, usually a quartz plate, is placed above the plasma in the TM-MW reactor to avoid overheating, thus allowing higher operating power (2–3 kW) and pressure (several tens of Torr or several thousand Pa). The key internal discharge parameters, such as the charged-particle concentrations and the electron energy distribution function

(EEDF), can be determined by Langmuir probe measurements. Experimental results showed that a high-density (~10^{12} cm^{-3}) plasma can be produced at an operating power on the order of 1 kW [14, 15].

In some cases, for both TE- and TM-MW reactors, a dc bias (several hundred volts) is applied to the growth substrate by introducing a parallel plate [16, 17] to promote the growth and alignment of VGs. The distance between the substrate and the parallel plate, as well as the in-series ballast-resistance, should be well adjusted to avoid the short circuit.

3.2.2 Radio Frequency Plasmas

Plasma driven at the RF domain is another popular power source for VG synthesis. There are three main modes to couple the energy of an RF generator to the plasmas: the evanescent electromagnetic (H) mode, the propagating wave (W) mode, and the electrostatic (E) mode [18].

H-mode ICP is based on the principle that the energy from the RF power is coupled by an inductive circuit element (typically a helical or spiral-like conductor) adjacent to or immersed inside the discharge region [19]. The inductive coil stimulates the magnetic field in ICP discharges. The magnetic field further induces a high-frequency vortex electric field concentric with the elements of the coil, which is able to provide breakdown and sustain the inductively coupled discharge. The non-conservative electric field is relatively low, so the ICP plasma discharges usually operate at low pressures, where the reduced electric field E/p is sufficient for ionization [18]. Generally, there are two main geometric designs for ICP reactors, i.e., the planar coil geometry and the cylindrical source tube with an expanding chamber. The former has been extensively used for VG synthesis, where 13.56 MHz RF energy is inductively coupled through a quartz window into the deposition chamber and a 3-turn planar coiled RF antenna [20], as schematically shown in Fig. 3.4c. Based on the latter geometry, W-mode helicon (whistler wave) plasma is obtained by adding a helicon antenna to launch propagating electromagnetic waves, i.e., B_0, as shown in Fig. 3.4d. Introducing a static magnetic field to an ICP-excited plasma will lead to a higher energy density and a larger plasma volume, and thus makes helicon reactors attractive for VG synthesis with a high growth rate. Typical magnetic field, RF frequency, and RF power used for helicon plasma-assisted VG synthesis were 10 mT, 13.56 MHz, and 1000 W, respectively [21].

E-mode CCP can be produced by a pair of parallel plane electrodes separated by a small distance, with one electrode being connected to the power supply while the other being grounded. Compared with ICP, the CCP with parallel-plane geometry has a simpler apparatus and a higher operating pressure. However, CCP as an independent plasma source for VG synthesis is commonly considered to be insufficient [22] because of the relatively low electron density and electron energy. The plasma density and electron temperature of CCP (Langmuir probe measurements showed that the typical electron density was 10^9–10^{10}, and

10^{11} cm^{-3} for high-frequency CCPs [23, 24]) are obviously lower than those of the above mentioned high-density plasma sources (10^{10}–10^{12} cm^{-3} for MW and ICP plasmas, and 10^{13} cm^{-3} for helicon plasmas) [18, 25]. Meanwhile, the high sheath potential of CCP could possibly destroy the surface bonds and prevent the growth of high quality crystals. Contamination from the electrode is another potential issue. To our knowledge, the only successful application of CCP as an independent plasma source for VG growth was demonstrated by combining an expanding RF plasma with a magnetron sputtering setup, as shown in Fig. 3.4g, where the electron density could reach the 10^{11} cm^{-3} level. In this case, the presence of a nanostructured catalyst is an indispensable factor for the successful growth of VG [26, 27].

As an alternative, CCP was used for VG synthesis together with other high-density plasma sources, such as CCP+ICP, and VHFCCP+MW, as shown in Fig. 3.4e, f, respectively. The basic principle of these combinations is that the hydrocarbon (e.g., CH_4) or fluorocarbon (e.g., CF_4 and C_2F_6) gases are initially excited by a parallel-plate CCP to form CH_x or/and CF_x radicals ($x = 1-3$), while remote supplemental H radicals are provided by high-density plasma sources [22]. This design takes advantage of the simultaneous formation of high density CH_x/CF_x radicals for large area synthesis and the sufficient production of H radicals for the removal of excess a-C. For a CCP+ICP VG synthesis system, typical RF frequencies and power for CCP and ICP were 13.56 MHz/100 W and 13.56 MHz/400 W, respectively [28]. By adding a remote ICP H radical source, the H radical density can be increased by several times, which benefits the VG synthesis and the morphology control [10]. Based on a similar principle, the combination of CCP at VHF and MW was also used for VG synthesis, where MW served as the high-intensity plasma source for H radical injection [29, 30]. CCPs at VHF can possibly lead to the transition from E- to H-mode, benefiting the electron density and the operating pressure. For VHFCCP+MW systems, typical frequencies and power for VHFCCP and MW were 100 MHz/300 W and 2.45 GHz/250 W, respectively [29].

3.2.3 Direct Current Plasmas

It is commonly accepted that the discovery of VG was first reported in 1997 when a petal-like carbon nanorose was obtained by dc arc discharge-assisted evaporation of graphite [13]. A VG network with improved alignment was synthesized by Obraztsov et al. [31] using parallel-plate dc glow PECVD. Later research on VG synthesis mainly focused on the dc glow discharges with different electrode arrangements.

Parallel-plate dc glow (schematically shown in Fig. 3.4h) has a simple setup and is commonly used in dc PECVD practice for VG synthesis. When a potential applied between the planar cathode and anode is sufficiently high, the so-called Townsend breakdown occurs. The minimum breakdown voltage (V_B) depends on a particular gas composition, pressure (*p*), and electrode distance (*d*), which can be expressed by the Paschen's law [32]:

$$V_B = \frac{a \cdot p \cdot d}{\ln(p \cdot d) + b}, \tag{3.1}$$

where *a* and *b* are constants that are mainly dependent on the gas. For a basic parallel-plate dc glow discharge, there are eight main regions distinguished from each other. Along the direction from the cathode to the anode, these eight regions are: Aston dark space, cathode glow, cathode dark space (cathode sheath), negative glow, Faraday space, positive column, anode glow, and anode dark space [25]. The substrate is usually placed on the top of the cathode (in some cases the substrate also serves as the cathode directly) since most of the applied potential difference drops in the first millimeters near the cathode, and the strong electric field and high ion flux within the cathode sheath are believed to benefit both the growth rate and the alignment of VG sheets during growth. For parallel-plate dc glow PECVD systems, the typical voltage and power for VG synthesis were −50 to −250 V and 3 kW, respectively; and the inter-electrode gap was several centimeters [33, 34].

In addition to the parallel-plate design, dc glow PECVD synthesis of VG was also demonstrated using a pin-to-plate electrode pair [7, 35]. As shown in Fig. 3.4i, a pair of asymmetric discharge electrodes, i.e., a sharpened tungsten tip and a planar substrate, are used in this PECVD system. Benefiting from the highly enhanced electric field generated near the tungsten tip, the VG growth in the pin-to-plate dc PECVD system can be operated at atmospheric pressure with a relatively high growth rate, showing the potential for massive production. However, the inherently non-uniform characteristic of the pin-to-plate glow discharge plasma resulted in non-uniformity in both the morphology and the alignment of as-grown VG sheets on the substrate. As suggested by Denysenko et al. [36] the effect of plasma non-uniformity can play a significant role in the PECVD process.

3.3 Raman Spectroscopy of VG

Raman spectroscopy is a fundamental and powerful tool for carbon material characterization, which has been widely used to determine the structural and electronic properties of VG nanosheets. In this section, we introduce some Raman spectroscopy results of the VG sheets obtained from different plasma sources. In a typical Raman spectrum of VG, as shown in Fig. 3.5a, three main peaks, namely the G, D, and D′ bands, appear at ~1580, ~1350, and ~1620 cm^{-1}, respectively. The G band is due to the stretch vibration mode of the C–C bond in graphitic materials and is commonly found in all sp^2 carbon structures. The D band corresponds with the disorder-induced phonon mode and the presence of disorders in the structure. The D′ band is related to the finite sp^2 crystallite size. Additionally, three second order peaks located at ~2652 cm^{-1} (G′ or 2D band), ~2915 cm^{-1} (D + G band), and ~3240 (2D′ band) were also found in the spectrum [37].

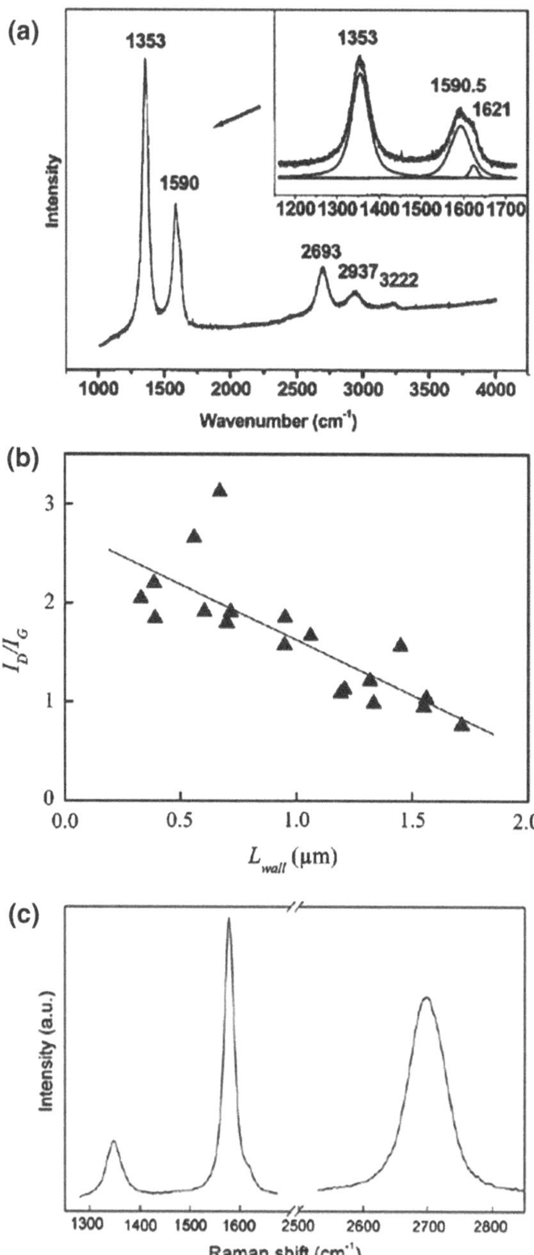

Fig. 3.5 **a** Raman spectrum of VG nanosheets excited by 514 nm laser; the *inset* shows the 1000–1800 cm^{-1} range, together with fitted peaks. Reprinted with permission from [37]. Copyright 2006 AIP Publishing LLC. **b** *I*(D)/*I*(G) as a function of the average length of VG nanosheets. Reprinted with permission from [33]. Copyright 2005 AIP Publishing LLC. **c** The Raman spectrum of VG presenting a low *I*(D)/*I*(G) ratio. Reprinted with permission from [3]. Copyright 2008 IOP Publishing

Most studies on Raman spectra of VG were focused on the shift and the full width at half maximum (FWHM) of G and G′ peaks (sometime includes D and D′ peaks as well [33]) as well as the intensity ratio of D peak-to-G peak *I*(D)/*I*(G). Generally, a higher *I*(D)/*I*(G) value and an increased FWHM value of the bands suggest smaller crystalline grains as well as a smaller inter-defect distance [38]. The FWHM of the G band was reported as in the range of 22.5–77.01 cm^{-1}: 22.5–60 cm^{-1} for TM-MW PECVD [38, 39], 25–55 cm^{-1} for TM-MW PECVD [40], 39.3–43.2 cm^{-1} for dc glow PECVD [33], and 40.28–77.01 cm^{-1} for dc glow PECVD [7]. Kurita et al. [33] suggested that the size of the as-grown VG can be estimated from the value of *I*(D)/*I*(G) since *I*(D)/*I*(G) decreases linearly with the lateral length of an individual VG as shown in Fig. 3.5b, which has been confirmed by many studies [7, 41–43]. This can be explained by the empirical equation proposed by Cancado et al. [44] which describes the relation among the in-plane sp^2 crystallite size L_a, the excitation energy of laser source E_L, and *I*(D)/*I*(G):

$$L_a\,(\mathrm{nm}) = \frac{560}{E_L^4}\left(\frac{I_D}{I_G}\right)^{-1}. \tag{3.2}$$

Some of the as-reported *I*(D)/*I*(G) values are: 0.2–0.3 for TM-MW PECVD (as shown in Fig. 3.5c, excitation 514 nm) [3], 0.3 for TM-MW PECVD (excitation 514.5 nm) [40, 43, 45], 0.35–1.11 for ICP (excitation 514 nm) [41, 46], 0.65–0.93 for TM-MW (excitation 633 nm) [38, 39], 0.77–2.66 for dc glow PECVD (excitation 532 nm) [33], 0.99 for TM-MW [47], 1.35–2.43 for TE-MW (excitation 514 nm) [37], and 2.39–3.28 for dc glow PECVD (excitation 633 nm laser) [7]. Details on the growth conditions of the above mentioned processes are listed in Table 3.1. It seems that high-intensity plasmas tend to produce VG nanosheets with a high degree of graphitization (relatively low *I*(D)/*I*(G) values). However, one should be very careful to make a general conclusion on which plasma source is the best choice with a simple comparison of *I*(D)/*I*(G) values. For example, the *I*(D)/*I*(G) value dramatically varies with the growth time even for the same plasma source, precursors, and operating parameters, and also depends on the Raman laser [7]. PECVD growth of VG is a complex process and the plasma source is only one of the critical parameters that determine the material quality. We will discuss other critical parameters in the VG synthesis in the following chapters.

3.4 Summary

PECVD, as a key method for VG synthesis, has several advantageous features, such as a low substrate temperature, high growth selectivity, and good control over nanostructure ordering/patterning. In this chapter, we first discussed the growth mechanism of VG during a typical PECVD process, which is generally recognized to comprise three consecutive steps: nucleation, growth, and termination. The nature of the plasma source and the operating parameters of a PECVD reactor are found to have significant impacts on the morphological and structural characteristics of VG sheets. We discussed three types of plasma sources (namely, MW, RF, and dc plasmas) and the different configurations and features of reactors based on those plasmas. We also briefly compared the structure of VG sheets produced from different plasma sources using Raman spectroscopy. The other important operating parameters of PECVD reactors will be discussed in Chap. 4.

References

1. Bo, Z., Yang, Y., Chen, J., Yu, K., Yan, J., & Cen, K. (2013). Plasma-enhanced chemical vapor deposition synthesis of vertically oriented graphene nanosheets. *Nanoscale, 5*(12), 5180–5204.
2. Bo, Z., Mao, S., Han, Z. J., Cen, K., Chen, J., & Ostrikov, K. (2015). Emerging energy and environmental applications of vertically-oriented graphenes. *Chemical Society Reviews*. doi:10.1039/C4CS00352G.
3. Malesevic, A., Vitchev, R., Schouteden, K., Volodin, A., Zhang, L., Van Tendeloo, G., et al. (2008). Synthesis of few-layer graphene via microwave plasma-enhanced chemical vapour deposition. *Nanotechnology, 19*(30), 305604.
4. Davami, K., Shaygan, M., Kheirabi, N., Zhao, J., Kovalenko, D. A., Rummeli, M. H., et al. (2014). Synthesis and characterization of carbon nanowalls on different substrates by radio frequency plasma enhanced chemical vapor deposition. *Carbon, 72*, 372–380.
5. Cai, M., Outlaw, R. A., Butler, S. M., & Miller, J. R. (2012). A high density of vertically-oriented graphenes for use in electric double layer capacitors. *Carbon, 50*(15), 5481–5488.
6. Zhao, J., Shaygan, M., Eckert, J., Meyyappan, M., & Rummeli, M. H. (2014). A growth mechanism for free-standing vertical graphene. *Nano Letters, 14*(6), 3064–3071.
7. Bo, Z., Yu, K., Lu, G., Wang, P., Mao, S., & Chen, J. (2011). Understanding growth of carbon nanowalls at atmospheric pressure using normal glow discharge plasma-enhanced chemical vapor deposition. *Carbon, 49*(6), 1849–1858.
8. Ostrikov, K., Neyts, E. C., & Meyyappan, M. (2013). Plasma nanoscience: From nano-solids in plasmas to nano-plasmas in solids. *Advances in Physics, 62*(2), 113–224.
9. Yu, K., Wang, P., Lu, G., Chen, K.-H., Bo, Z., & Chen, J. (2011). Patterning vertically oriented graphene sheets for nanodevice applications. *Journal of Physical Chemistry Letters, 2*(6), 537–542.
10. Hiramatsu, M., Shiji, K., Amano, H., & Hori, M. (2004). Fabrication of vertically aligned carbon nanowalls using capacitively coupled plasma-enhanced chemical vapor deposition assisted by hydrogen radical injection. *Applied Physics Letters, 84*(23), 4708–4710.
11. Zhu, M., Wang, J., Holloway, B. C., Outlaw, R. A., Zhao, X., Hou, K., et al. (2007). A mechanism for carbon nanosheet formation. *Carbon, 45*(11), 2229–2234.
12. Seo, D. H., Rider, A. E., Han, Z. J., Kumar, S., & Ostrikov, K. (2013). Plasma break-down and re-build: Same functional vertical graphenes from diverse natural precursors. *Advanced Materials, 25*(39), 5638–5642.
13. Ando, Y., Zhao, X., & Ohkohchi, M. (1997). Production of petal-like graphite sheets by hydrogen arc discharge. *Carbon, 35*(1), 153–158.
14. Sugai, H., Ghanashev, I., & Mizuno, K. (2000). Transition of electron heating mode in a planar microwave discharge at low pressures. *Applied Physics Letters, 77*(22), 3523–3525.
15. Nagatsu, M., Xu, G., Ghanashev, I., Kanoh, M., & Sugai, H. (1997). Mode identification of surface waves excited in a planar microwave discharge. *Plasma Sources Science and Technology, 6*(3), 427–434.
16. Wu, Y. H., Qiao, P. W., Chong, T. C., & Shen, Z. X. (2002). Carbon nanowalls grown by microwave plasma enhanced chemical vapor deposition. *Advanced Materials, 14*(1), 64–67.

17. Zhang, Y., Du, J. L., Tang, S., Liu, P., Deng, S. Z., Chen, J., & Xu, N. S. (2012). Optimize the field emission character of a vertical few-layer graphene sheet by manipulating the morphology. *Nanotechnology, 23*(1), 015202.
18. Chabert, P. & Braithwaite, N. (2001). *Physics of radio-frequency plasmas* (pp. 1–385). New York: Cambridge University Press.
19. Hopwood, J. (1992). Review of inductively coupled plasmas for plasma processing. *Plasma Sources Science and Technology, 1*(2), 109–116.
20. Wang, J. J., Zhu, M. Y., Outlaw, R. A., Zhao, X., Manos, D. M., & Holloway, B. C. (2004). Synthesis of carbon nanosheets by inductively coupled radio-frequency plasma enhanced chemical vapor deposition. *Carbon, 42*(14), 2867–2872.
21. Sato, G., T. Morio, T. Kato, & R. Hatakeyama. (2006). Fast growth of carbon nanowalls from pure methane using helicon plasma-enhanced chemical vapor deposition. *Japanese Journal of Applied Physics Part 1—Regular Papers Brief Communications & Review Papers, 45*(6A), 5210–5212.
22. Ostrikov, K., Cvelbar, U., & Murphy, A. B. (2011). Plasma nanoscience: Setting directions, tackling grand challenges. *Journal of Physics D-Applied Physics, 44*(17), 174001.
23. Paranjpe, A. P., McVittie, J. P., & Self, S. A. (1990). A tuned langmuir probe for measurements in RF glow-discharges. *Journal of Applied Physics, 67*(11), 6718–6727.
24. Hopwood, J., Guarnieri, C. R., Whitehair, S. J., & Cuomo, J. J. (1993). Langmuir probe measurements of a radio-frequency induction plasma. *Journal of Vacuum Science and Technology a-Vacuum Surfaces and Films, 11*(1), 152–156.
25. Lieberman, M. A. & Lichtenberg, A. J. (2005). *Principles of plasma discharges and materials processing* (2nd ed., pp. 1–757). New Jersey: Wiley.
26. Vizireanu, S., Stoica, S. D., Luculescu, C., Nistor, L. C., Mitu, B., & Dinescu, G. (2010). Plasma techniques for nanostructured carbon materials synthesis. A case study: Carbon nanowall growth by low pressure expanding RF plasma. *Plasma Sources Science and Technology, 19*(3), 034016.
27. Malesevic, A., Vizireanu, S., Kemps, R., Vanhulsel, A., Van Haesendonck, C., & Dinescu, G. (2007). Combined growth of carbon nanotubes and carbon nanowalls by plasma-enhanced chemical vapor deposition. *Carbon, 45*(15), 2932–2937.
28. Shiji, K., Hiramatsu, M., Enomoto, A., Nakamura, N., Amano, H., & Hori, M. (2005). Vertical growth of carbon nanowalls using rf plasma-enhanced chemical vapor deposition. *Diamond and Related Materials, 14*(3–7), 831–834.
29. Kondo, S., Hori, M., Yamakawa, K., Den, S., Kano, H., & Hiramatsu, M. (2008). Highly reliable growth process of carbon nanowalls using radical injection plasma-enhanced chemical vapor deposition. *Journal of Vacuum Science and Technology B, 26*(4), 1294–1300.
30. Takeuchi, W., Ura, M., Hiramatsu, M., Tokuda, Y., Kano, H., & Hori, M. (2008). Electrical conduction control of carbon nanowalls. *Applied Physics Letters, 92*(21), 213103.
31. Obraztsov, A. N., Volkov, A. P., Nagovitsyn, K. S., Nishimura, K., Morisawa, K., Nakano, Y., & Hiraki, A. (2002). CVD growth and field emission properties of nanostructured carbon films. *Journal of Physics D-Applied Physics, 35*(4), 357–362.
32. Paschen, F. (1889). Ueber die zum Funkenübergang in Luft, Wasserstoff und Kohlensäure bei verschiedenen Drucken erforderliche Potentialdifferenz. *Annalen der Physik, 273*(5), 69–96.
33. Kurita, S., Yoshimura, A., Kawamoto, H., Uchida, T., Kojima, K., Tachibana, M., et al. (2005). Raman spectra of carbon nanowalls grown by plasma-enhanced chemical vapor deposition. *Journal of Applied Physics, 97*(10), 104320.
34. Banerjee, D., Mukherjee, S., & Chattopadhyay, K. K. (2011). Synthesis of amorphous carbon nanowalls by DC-PECVD on different substrates and study of its field emission properties. *Applied Surface Science, 257*(8), 3717–3722.
35. Yu, K., Bo, Z., Lu, G., Mao, S., Cui, S., Zhu, Y., et al. (2011). Growth of carbon nanowalls at atmospheric pressure for one-step gas sensor fabrication. *Nanoscale Research Letters, 6*, 202.
36. Denysenko, I. B., Xu, S., Long, J. D., Rutkevych, P. P., Azarenkov, N. A., & Ostrikov, K. (2004). Inductively coupled Ar/CH4/H-2 plasmas for low-temperature deposition of ordered carbon nanostructures. *Journal of Applied Physics, 95*(5), 2713–2724.

37. Ni, Z. H., Fan, H. M., Feng, Y. P., Shen, Z. X., Yang, B. J., & Wu, Y. H. (2006). Raman spectroscopic investigation of carbon nanowalls. *Journal of Chemical Physics, 124*(20), 204703.
38. Soin, N., Roy, S. S., O'Kane, C., McLaughlin, J. A. D., Lim, T. H., & Hetherington, C. J. D. (2011). Exploring the fundamental effects of deposition time on the microstructure of graphene nanoflakes by Raman scattering and X-ray diffraction. *CrystEngComm, 13*(1), 312–318.
39. Soin, N., Roy, S. S., Lim, T. H., & McLaughlin, J. A. D. (2011). Microstructural and electrochemical properties of vertically aligned few layered graphene (FLG) nanoflakes and their application in methanol oxidation. *Materials Chemistry and Physics, 129*(3), 1051–1057.
40. Teii, K., Shimada, S., Nakashima, M., & Chuang, A. T. H. (2009). Synthesis and electrical characterization of n-type carbon nanowalls. *Journal of Applied Physics, 106*(8), 084303.
41. Zhu, M. Y., Outlaw, R. A., Bagge-Hansen, M., Chen, H. J., & Manos, D. M. (2011). Enhanced field emission of vertically oriented carbon nanosheets synthesized by C2H2/H-2 plasma enhanced CVD. *Carbon, 49*(7), 2526–2531.
42. Jain, H. G., Karacuban, H., Krix, D., Becker, H.-W., Nienhaus, H., & Buck, V. (2011). Carbon nanowalls deposited by inductively coupled plasma enhanced chemical vapor deposition using aluminum acetylacetonate as precursor. *Carbon, 49*(15), 4987–4995.
43. Cheng, C. Y., & Teii, K. (2012). Control of the growth regimes of nanodiamond and nanographite in microwave plasmas. *IEEE Transactions on Plasma Science, 40*(7), 1783–1788.
44. Cancado, L. G., Takai, K., Enoki, T., Endo, M., Kim, Y. A., Mizusaki, H., et al. (2006). General equation for the determination of the crystallite size L-a of nanographite by Raman spectroscopy. *Applied Physics Letters, 88*(16), 163106.
45. Teii, K., & Ikeda, T. (2007). Effect of enhanced C-2 growth chemistry on nanodiamond film deposition. *Applied Physics Letters, 90*(11), 111504.
46. French, B. L., Wang, J. J., Zhu, M. Y., & Holloway, B. C. (2005). Structural characterization of carbon nanosheets via X-ray scattering. *Journal of Applied Physics, 97*(11), 114317.
47. Chuang, A. T. H., Boskovic, B. O., & Robertson, J. (2006). Freestanding carbon nanowalls by microwave plasma-enhanced chemical vapour deposition. *Diamond and Related Materials, 15*(4–8), 1103–1106.
48. Wu, Y. H., Yu, T., & Shen, Z. X. (2010). Two-dimensional carbon nanostructures: Fundamental properties, synthesis, characterization, and potential applications. *Journal of Applied Physics, 108*(7), 071301.
49. Wang, Z., Shoji, M., & Ogata, H. (2011). Carbon nanosheets by microwave plasma enhanced chemical vapor deposition in CH4-Ar system. *Applied Surface Science, 257*(21), 9082–9085.
50. Shang, N. G., Papakonstantinou, P., McMullan, M., Chu, M., Stamboulis, A., Potenza, A., et al. (2008). Catalyst-free efficient growth, orientation and biosensing properties of multilayer graphene nanoflake films with sharp edge planes. *Advanced Functional Materials, 18*(21), 3506–3514.
51. Mori, T., Hiramatsu, M., Yamakawa, K., Takeda, K., & Hori, M. (2008). Fabrication of carbon nanowalls using electron beam excited plasma-enhanced chemical vapor deposition. *Diamond and Related Materials, 17*(7–10), 1513–1517.
52. Zeng, L., Lei, D., Wang, W., Liang, J., Wang, Z., Yao, N., & Zhang, B. (2008). Preparation of carbon nanosheets deposited on carbon nanotubes by microwave plasma-enhanced chemical vapor deposition method. *Applied Surface Science, 254*(6), 1700–1704.
53. Chatei, H., Belmahi, M., Assouar, M. B., Le Brizoual, L., Bourson, P., & Bougdira, J. (2006). Growth and characterisation of carbon nanostructures obtained by MPACVD system using CH_4/CO_2 gas mixture. *Diamond and Related Materials, 15*(4–8), 1041–1046.
54. Obraztsov, A. N., Zolotukhin, A. A., Ustinov, A. O., Volkov, A. P., Svirko, Y., & Jefimovs, K. (2003). DC discharge plasma studies for nanostructured carbon CVD. *Diamond and Related Materials, 12*(3–7), 917–920.
55. Jiang, N., Wang, H. X., Zhang, H., Sasaoka, H., & Nishimura, K. (2010). Characterization and surface modification of carbon nanowalls. *Journal of Materials Chemistry, 20*(24), 5070–5073.
56. Krivchenko, V. A., Dvorkin, V. V., Dzbanovsky, N. N., Timofeyev, M. A., Stepanov, A. S., Rakhimov, A. T., et al. (2012). Evolution of carbon film structure during its catalyst-free growth in the plasma of direct current glow discharge. *Carbon, 50*(4), 1477–1487.

Chapter 4
PECVD Synthesis of Vertically-Oriented Graphene: Precursor and Temperature Effects

Abstract The plasma-enhanced chemical vapor deposition (PECVD) is an effective method for producing vertically-oriented graphene (VG) sheets, but it is also a very complex one because of the complexity associated with the plasma chemistry. In addition to the type of plasma sources discussed in Chap. 3, the morphology and structure of the PECVD-produced VG sheets are also strongly affected by a set of operating parameters, including precursors (e.g., feedstock gas type and composition, plasma gas type), the substrate temperature, and the operating pressure. In this chapter, we discuss two important operating parameters for the synthesis of VG in the PECVD process, specifically precursor and temperature.

Keywords Amorphous carbon · Argon · Carbon source · Etchant · Feedstock gas · Gas proportion · Graphene · Nitrogen · Orientation · Plasma-enhanced chemical vapor deposition · Precursor · Specific surface area · Temperature · Vertically-oriented graphene

4.1 Precursors

With a few exceptions of using metal–organic precursors [e.g., evaporated aluminum acetylacetonate ($Al(acac)_3$)] [1], most plasma-enhanced chemical vapor deposition (PECVD) systems use gaseous carbon-containing species (mostly hydrocarbons and fluorocarbons) as the carbon source for vertically-oriented graphene (VG) growth [2]. The feedstock gases of most PECVD processes for VG synthesis are listed in Table 3.1, where the corresponding operating conditions are also presented. Extensive studies have revealed that the feedstock gas composition and proportion significantly affect plasma property, synthesis process, as well as

---

Part of this chapter was adapted from our review article "Plasma-Enhanced Chemical Vapor Deposition Synthesis of Vertically-Oriented Graphene Nanosheets," Nanoscale 5(12), 5180–5204, 2013 (DOI: 10.1039/C3NR33449J)—Reproduced by permission of The Royal Society of Chemistry.

J. Chen et al., *Vertically-Oriented Graphene*, DOI 10.1007/978-3-319-15302-5_4

the morphology and structure of the as-grown deposits [3–8]. In this section, we will first discuss the roles of each feedstock gas component in the synthesis process, and then present the effect of the feedstock gas proportion on plasma chemistry and VG growth.

4.1.1 Carbon Sources: Feedstock Gases

Hydrocarbon (C_2H_2 or CH_4) and fluorocarbon (CF_4, CHF_3, or C_2F_6) were the popular choices of carbon sources for most PECVD systems. Other particular cases mainly include the following: Mori et al. [9] performed VG growth in a TE-MW system with a CO/H_2 precursor, and Chatei et al. [10] conducted the growth of VG using a TE-MW reactor with a mixture of CH_4/CO_2 where both CH_4 and CO_2 could be carbon sources simultaneously.

CH_x ($x = 1–3$) radicals were believed to play an important role in the VG synthesis. Shiji et al. [11] conducted comparative work on the VG synthesis in a CCP + ICP system with CH_4 and a series of fluorocarbons diluted by H_2, and they found that VG was successfully synthesized when CH_4, CF_4, CHF_3, or C_2F_6 was used as the carbon source, but VG was not obtained using the C_4F_8/H_2 precursor. They further compared the growth rate and the VG morphology, including thickness and intersheet spacing, for different carbon sources. The main observations included the following: VG grown using the CH_4/H_2 system was rather wavy, while VG obtained using fluorocarbon/H_2 precursors showed maze-like morphology, as shown in Fig. 4.1a; the interlayer spacing was in the order of $CF_4 > CHF_3 > C_2F_6 > CH_4$, from large to small, as shown in Fig. 4.1a–d; and the growth rate was in the order of $C_2F_6 > CHF_3 > CH_4 > CF_4$, from high to low, as shown in Fig. 4.2. It is worth noting that the as-reported growth rate of VG (~0.18 μm/h for C_2F_6/H_2, ~0.16 μm/h for CHF_3/H_2, ~0.15 μm/h for CH_4/H_2, and ~0.11 μm/h for CF_4/H_2) is much lower than those obtained in TM-MW (~96 μm/h for CH_4/N_2) [12], helicon (~18 μm/h for CH_4) [4], ICP (~16 μm/h for C_2H_2, ~10 μm/h for C_2H_2/H_2, and ~2 μm/h for CH_4/H_2) [13, 14], TE-MW (~15 μm/h CH_4/H_2) [8], and dc glow (~1.5 μm/h for CH_4/H_2) [15] systems.

Obraztsov et al. [16] suggested that the presence of reactive carbon dimers (C_2) could play a crucial role in VG growth. Figure 4.3 shows the optical emission spectroscopy (OES) measurement results obtained in a dc glow plasma system employing pure H_2 and CH_4/H_2 mixtures. It was found that recombination lines of the atomic (656 nm, H_α; 487 nm, H_β) and molecular (550–650 nm, H_2) hydrogen dominated the emission spectrum when pure hydrogen was used as the feed gas, while the characteristic emission lines of CH radicals (390 and 430 nm) and C_2 dimers (515 and 560 nm) were obviously observed, especially at areas near the growth substrate. Based on the OES results and the empirical calculations proposed by Gruen [17], Obraztsov et al. [16] deduced the significant role of C_2 in the evolution of critical nuclei into VG sheets with the insertion of C_2 into acetylene-like C=C bond to produce a carbene structure.

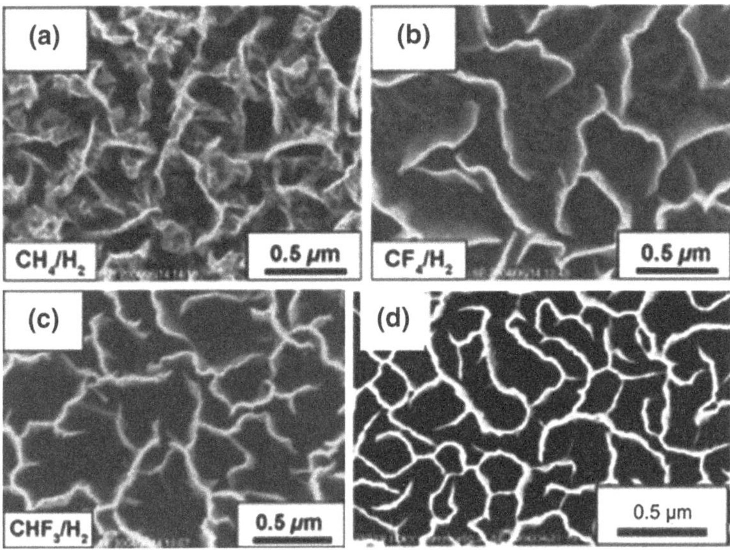

Fig. 4.1 SEM images of VG grown for 8 h using **a** the CH_4/H_2 system, **b** the CF_4/H_2 system, **c** the CHF_3/H_2 system, and **d** the C_2F_6/H_2 system. Reprinted with permission from [11]. Copyright 2005 Elsevier. Information on plasma type and growth conditions can be found in Table 3.1

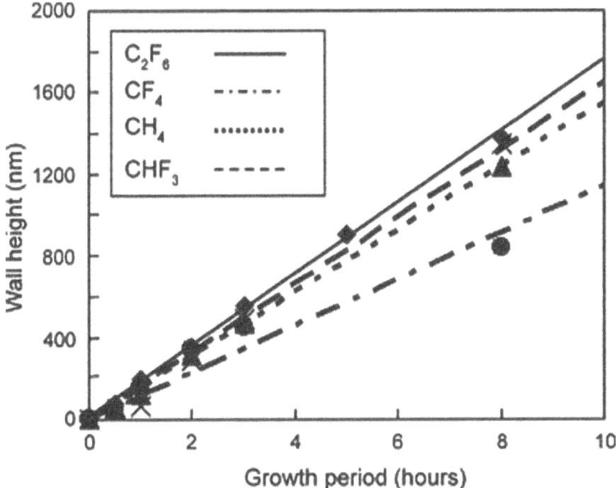

Fig. 4.2 VG growth height as a function of growth time for systems using C_2F_6 (*solid line*), CF_4 (*dash-dot line*), CH_4 (*dotted line*), and CHF_3 (*broken line*) as the carbon source gas. Reprinted with permission from [11]. Copyright 2005 Elsevier. Information on plasma type and growth conditions can be found in Table 3.1

The important role of C_2 in VG growth, especially for the formation of critical nuclei, was mentioned by many groups [6, 7, 10, 13, 18]. C_2 radical density of 10^{11}–10^{13} cm^{-3} was demonstrated in the MW and ICP systems employing a CH_4/H_2 or $CH_4/H_2/Ar$ mixture [19–21]. Teii et al. [6] conducted plasma diagnosis and

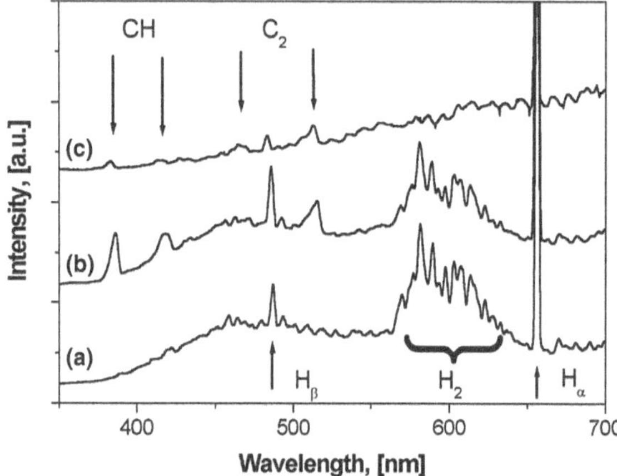

Fig. 4.3 Typical OES for pure hydrogen (**a**) and for a hydrogen–methane gas mixture with 10 % (**b**) and 25 % (**c**) of methane. Total gas pressure is 80 Torr (~1×10^4 Pa), and applied voltage is 650 V (**a**), 750 V (**b**), and 850 V (**c**). The discharge current is 7 A (**a**), 6 A (**b**), and 5 A (**c**). Reprinted with permission from [17]. Copyright 2003 Elsevier. Information on plasma type and growth conditions can be found in Table 3.1

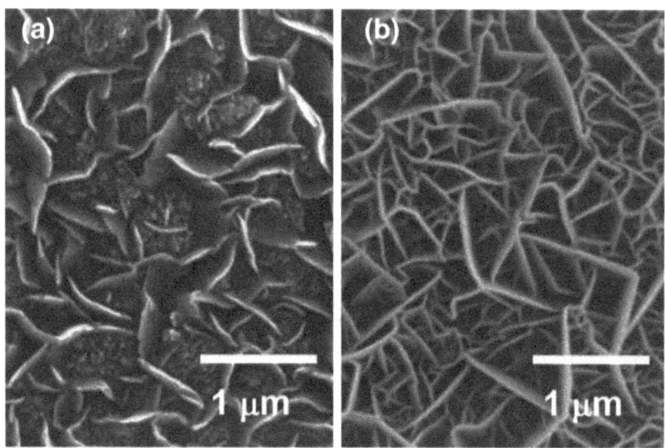

Fig. 4.4 SEM images of the deposits obtained on Si substrates in **a** $CH_4/N_2/Ar$ (4:10:86) at a deposition temperature of 1150 °C and **b** $C_2H_2/N_2/Ar$ (0.5:29.5:70) at a deposition temperature of 950 °C. Reprinted with permission from [6]. Copyright 2009 AIP Publishing LLC. Information on plasma type and growth conditions can be found in Table 3.1

material characterization on a TM-MW plasma (employing $C_2H_2/N_2/Ar$ or $CH_4/N_2/Ar$) as well as the as-grown deposits. As shown in Fig. 4.4a, b, the deposits from $C_2H_2/N_2/Ar$ were networks of VG, while the deposits from $CH_4/N_2/Ar$ consisted of VG intercepted by diamonds, which was attributed to the different C_2 density in these two precursor systems. They further proposed different pathways for C_2H_2 and CH_4 in their transformation to C_2; that is, C_2H_2 was able to produce C_2 through

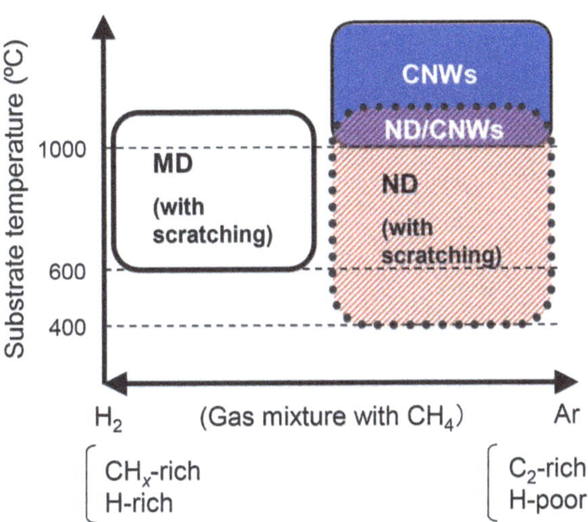

Fig. 4.5 Variation of the deposits as a function of substrate temperature and gas composition. *MD* microcrystalline diamond; *ND* nanodiamond; *CNWs* carbon nanowalls (VG). Reprinted with permission from [33]. Copyright 2012 IEEE

direct dissociation due to the strong C≡C bond, while CH_4 was easily converted to CH_x ($x = 1$–3) radicals to produce C_2 through radical recombination and subsequent dissociation. This explained the absence or near absence of diamonds in the deposits of $C_2H_2/N_2/Ar$ system, where a higher C_2 radical density facilitated the formation of non-diamond carbon components, including microcrystals of graphite, carbides, and a-C [22].

Very recently, Teii et al. proposed the growth regimes for nanodiamond and VG, as shown in Fig. 4.5, and concluded that VG grew from C_2 and unsaturated radicals. However, the as-proposed growth regime of VG is not a universal one since the growth of VG without the formation of diamond at a lower substrate temperature has been widely demonstrated using other systems (see Table 3.1) [3, 4, 6, 8, 10, 11, 13, 14, 23–32].

Zhu et al. [13, 31] compared the growth of VG in an ICP system employing CH_4/H_2 and C_2H_2/H_2 precursors. VG growth using C_2H_2/H_2 ICP had a higher growth rate, and the as-grown VG sheets presented more ordered vertical orientation and more uniform sheet height distribution, compared with the CH_4/H_2 counterpart, as shown in the top view and side view SEM images in Fig. 4.6a–d. Meanwhile, the edge thickness of a single VG nanosheet obtained by C_2H_2/H_2 growth was about 1–2 nm (3–4 atomic layers as shown in the TEM image of Fig. 4.6e), which was slightly larger than that of CH_4/H_2 growth (<1 nm as shown in the SEM image of Fig. 4.7). The growth temperature could be lowered by 100–150 °C when using C_2H_2 instead of CH_4 as a feedstock. They attributed the above observation to the higher carbon-bearing species density in the plasmas employing C_2H_2 feedstock due to the lower dissociation energy. It is worth noting that the use of C_2H_2 could possibly induce a short-circuit risk for some PECVD systems due to the easy production and deposition of a-C in an uncontrolled manner.

Vizireanu et al. [18] suggested that ionic clusters of carbon $C_nH_x^+$ ($n \geq 2$, $x = 1$, 2, 3) are the building species for VG nanosheet growth, based on their OES and

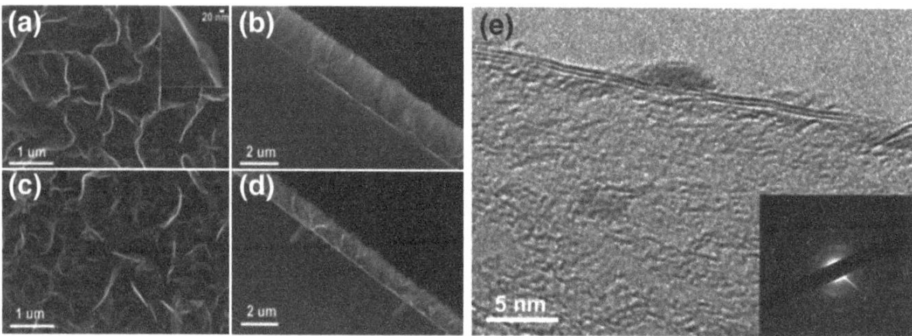

Fig. 4.6 **a** Top view SEM image of VG nanosheets deposited on a Si substrate using 80 % C_2H_2 in H_2, at 600 °C substrate temperature, 35 m Torr (4.66 Pa) total pressure, and 1000 W RF power for 10 min. *Inset* enlarged SEM image shows the edge thickness of 1–2 nm. **b** Side view of the C_2H_2 VG nanosheets shown in (**a**). **c** Top view of typical (40 % CH_4 in H_2, 700 °C, 100 m Torr (13.3 Pa), and 900 W) CH_4 VG nanosheets deposited on a Si substrate for 20 min. **d** Side view of the CH_4 VG nanosheets shown in (**c**). **e** High-resolution TEM (HRTEM) image of C_2H_2 VG nanosheets grown directly on a Cu grid. *Inset* selected area electron diffraction pattern of the same nanosheet. Reprinted with permission from [13]. Copyright 2011 Elsevier

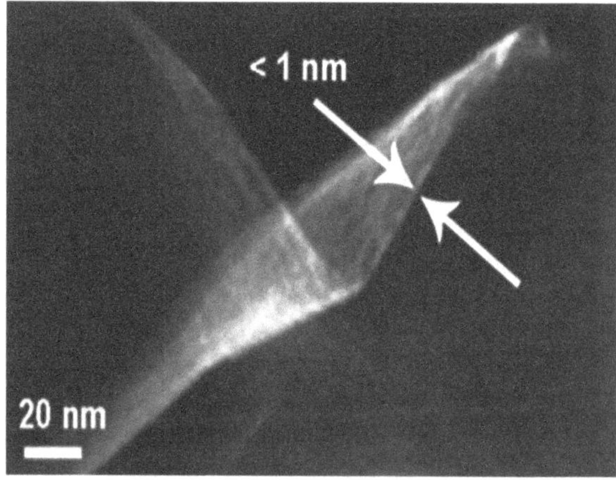

Fig. 4.7 SEM image of an enlarged CH_4 VG nanosheet. Reprinted with permission from [31]. Copyright 2007 Elsevier. Information on plasma type and growth conditions can be found in Table 3.1

mass spectroscopy measurements on an RF expanding plasma and the morphology characterization of the as-grown VG nanosheets. As shown in Fig. 3.2g, the employed RF expanding plasma allowed the easy adjustment of the distance between the substrate and the gas injection point, and thus facilitated the investigation on the plasma chemistry involving both the species transported from the interelectrode space and those locally generated. Figure 4.8a shows the mass spectrum of the ions sampled from plasma, recorded at 5 cm from the injection point, where the quality of the VG nanosheets was considered as the highest, compared with those obtained closer or farther from injection. Correspondingly, as shown in Fig. 4.8b, although generally the intensities of C_2, CH, H, and Ar

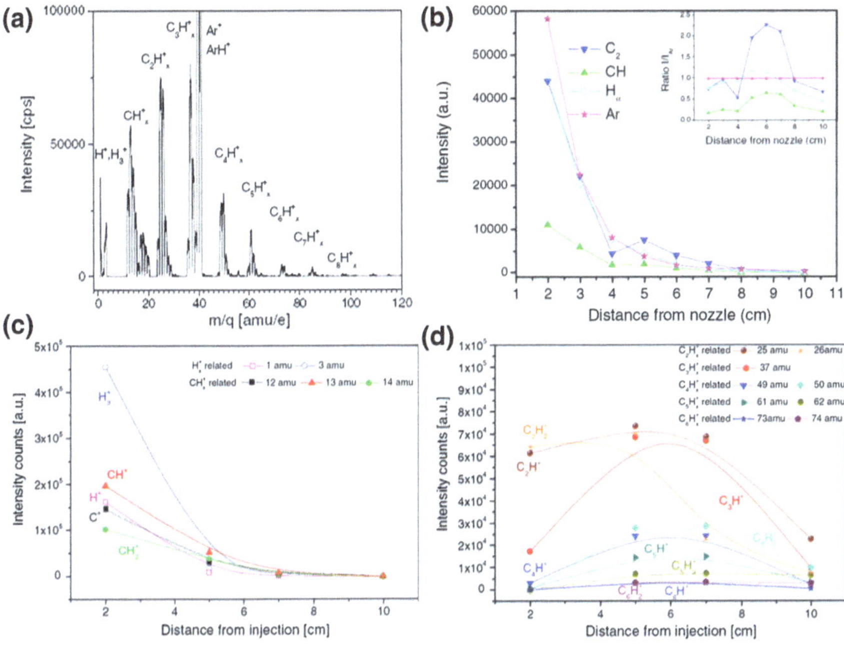

Fig. 4.8 **a** The mass spectrum of the ions sampled from plasma, recorded at 5 cm from the injection point. **b** Dependence of emission intensities of C_2, CH, H, and Ar species as a function of distance; *inset* dependence upon the position of the ratio of local C_2, CH, and H intensities to the local intensity of Ar line. The dependence of the concentration of various ionic species upon the distance from the injection. **c** H_x^+ and CH_x^+ species and **d** ionic clusters of carbon $C_nH_x^+$ ($n \geq 2$, $x = 1, 2, 3$). Reprinted with permission from [18]. Copyright 2010 IOP Publishing. Information on plasma type and growth conditions can be found in Table 3.1

species decreased along the flow with an increasing distance from the injection, it was observed that there was an enhancement of radical emission at the region around 5 cm (see the inset of Fig. 4.8b). The mass spectrum intensities of H_x^+ and CH_x^+ species decreased along the flow, but those of ionic clusters of carbon $C_nH_x^+$ ($n \geq 2$, $x = 1, 2, 3$) presented a maximum value around 5 cm, as shown in Fig. 4.8c, d. The electron density and plasma potential were measured by a Langmuir probe as 3.5×10^{17} m^{-3} and 43 V, respectively. They further proposed a possible mechanism for the formation of a high-mass ionic cluster of carbon, including the formation of low-mass ions from the reactions induced by background Ar and H_2 ions/molecules as well as energetic electrons, and the ionic chain polymerization process.

In the CO/H_2 TE-MW system used by Mori et al. [9], a strong C_2 Swan band and a Balmer α line of atomic hydrogen were observed in their OES measurements, both of which were considered to be critical to the growth of VG. They suggested the important role of C_2 in the VG growth and further deduced a possible C_2 formation pathway involving C, C_2O, and CO.

4.1.2 Amorphous Carbon Etchants

Although some details of the complex plasma chemistry in the VG growth process have yet to be fully revealed, it has been widely accepted that a basic VG growth evolution consists of the initial nucleation, the subsequent formation of VG nanosheets from the nuclei, and the further vertical growth [5, 26, 31, 34]. The effective removal of a-C has been widely recognized as a crucial and inevitable step to the formation of high-quality nanoislands during the initial nucleation stage.

Possible a-C removal species include atomic hydrogen, excited nitrogen species, atomic oxygen, and hydroxyl radicals. These species have been reported by different groups using various precursors and plasma sources. Hiramatsu et al. used CCP + ICP and VHFCCP + MW systems with a mixture of hydrocarbon/ H_2 or fluorocarbon/H_2 to grow VG, and their work suggested that H atoms play the most significant role in the removal of undesirable amorphous phases. H atoms were also believed to extract bonding atoms from the gas-phase radicals migrating on the growing surface or edge of graphene layers (e.g., F from CF_x radicals when fluorocarbon/H_2 was used as the precursor) [11, 24, 25, 35]. Similarly, Zhu et al. [31] explored the growth mechanism for VG synthesis by ICP and found that H atoms acted as an effective etchant to rapidly remove a-C defects due to the different etching rates of H atoms to a-C, sp^2, and sp^3 hybridized carbon, benefiting both the crystalline graphitic structure and the sharp edges in the growth layers. Chuang et al. [5] used a TM-MW reactor to synthesize freestanding VG, with a mixture of C_2H_2/NH_3, where NH_3 served as the H atom source for a-C removal. Shang et al. [12] reported the growth of VG using a TM-MW system with a mixture of CH_4/N_2, and they suggested the advantage of nitrogen in etching a-C phase compared with hydrogen. Kondo et al. [34] investigated the initial growth process of VG for the conditions with and without an O_2 gas addition to a C_2F_6/H_2 VHFCCP + MW system, and they found an O_2 addition has the effect of reducing a-C and controlling VG nucleation. Chatei et al. [10] compared the as-grown VG using CH_4/CO_2 or CH_4/H_2 in a TE-MW reactor and concluded that OH radicals and O atoms have a stronger ability to etch a-C than H atoms. We applied a dc pin-to-plate negative normal glow PECVD reactor for the synthesis of VG under atmospheric pressure, and we found that OH radicals with an adequate concentration would benefit the etching of a-C and the growth of VG [26].

With the use of a high-intensity plasma source such as helicon, ICP, or TM-MW, VG growth can be realized using pure CH_4 as the feedstock gas, where CH_4 serves as the carbon source and the a-C etchant simultaneously. For example, Sato et al. [4] used a helicon discharge capable of producing plasmas with high-concentration electrons (~10^{11} cm^{-3}, according to Langmuir probe measurements, as shown in Fig. 4.9a) and highly energized ions, to synthesize VG using CH_4 without dilution. As shown in Fig. 4.9b, the strong optical emission lines of 656.2 nm of H_α, 486.1 nm of H_β, and 431.5 nm of CH were observed in the OES measurement on the CH_4 helicon plasma for 3 mT magnetic field, indicating both

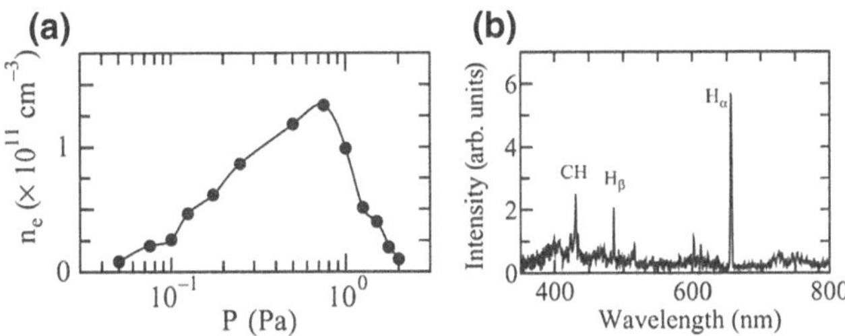

Fig. 4.9 **a** Plasma density dependence on gas pressure for 3 mT magnetic field. **b** Typical visible emissions from CH_4 helicon plasma for 3 mT magnetic field. Reprinted with permission from [4]. Copyright 2006 The Japan Society of Applied Physics. Information on plasma type and growth conditions can be found in Table 3.1

the large volume and highly energetic H atoms. Wang et al. [14] demonstrated VG growth in an ICP system using pure CH_4 as the precursor. They further conducted OES measurements on the plasmas produced by ICP and CCP reactors and found a dramatic increase in the intensity of H atom peaks from the ICP reactor relative to the CCP reactor [36]. The above practice employing relatively high-intensity plasmas ensures the sufficient formation of H atoms from the dissociation of CH_4 (without dilution) and allows for the efficient a-C removal for further growth of VG, where the pure hydrocarbon serves as the carbon source and the a-C etchant simultaneously. Besides the above practices using pure carbon source as the precursor, successful VG growth was also demonstrated in ICP and TM-MW systems using a mixture of carbon source gas and argon as the feedstock gases [27, 37]. Argon was not considered as an a-C etchant but has the ability to enhance the electron energy and favors the plasma stability. The role of argon in the PECVD process is discussed in detail in Sect. 4.1.3.

Besides the above-mentioned PECVD processes that employ pure hydrocarbon as the precursor, a more common way to ensure the sufficient production of a-C removal species was via introducing additional a-C etchant sources, such as hydrogen-rich gases (e.g., H_2 and NH_3) [3, 5–8, 11, 12, 14, 16, 25, 28–32], oxygen-rich gases (e.g., O_2 and CO_2) [10, 34], and hydroxyl-rich gas (e.g., H_2O) [26], into the precursor. These gas additions are especially desired for CCP or dc plasmas of relatively low intensity, where the shortage of a-C removal species from the direct dissociation of carbon sources is obvious. We conducted OES measurements on an atmospheric dc pin-to-plate negative normal glow PECVD reactor in CH_4/Ar, CH_4/Ar/H_2, and CH_4/Ar/H_2O, and compared the corresponding as-grown deposits [26]. As shown in Fig. 4.10a, only emissions from CH radicals (314 nm, $C^2\Sigma^{+a}$–$X^2\Pi$; 388 nm, $B^2\Sigma^{-a}$–$X^2\Pi$; 431 nm, $A^2\Delta$–$X^2\Pi$), C_2 radicals (516 nm, $A^3\Pi_g$–$X^3\Pi_u$), and Ar atoms (697, 707, 727, 738, 750, 764, 772, and 795 nm) were identified for the CH_4/Ar discharge, and there was only carbon black, but no VG was produced. By adding H_2 into the system (40 % H_2 by volume), a Balmer transition of H atoms (656 nm, H_α; 487 nm, H_β; 434 nm, H_γ) was observed, as shown in Fig. 4.10b, and accordingly, VG was observed in the as-grown deposits but with

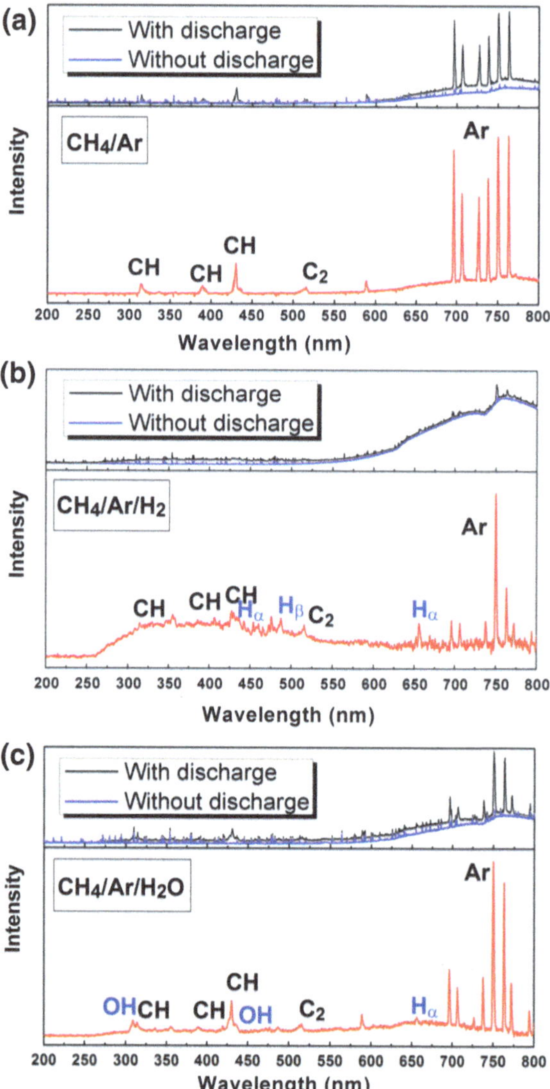

Fig. 4.10 Optical emission spectra for the normal glow discharge systems of **a** CH_4/Ar, **b** CH_4/Ar/H_2, and **c** CH_4/Ar/H_2O at a supply voltage of 3.35 kV and an interelectrode gap of 10 mm. The lower spectrum (*red*) in each panel was obtained by subtracting the background signal without discharge (*blue*) from the total spectrum with discharge (*black*) to eliminate the luminous emission contribution from the heated furnace. Reprinted with permission from [26]. Copyright 2011 Elsevier. Information on plasma type and growth conditions can be found in Table 3.1

a quite low growth rate. The above observation was attributed to the shortage of H atoms due to the relatively short mean free path of electrons or low electron energy in the atmospheric glow discharge. With H_2O addition, emissions from OH radicals (309 nm, $A^2\Sigma^+$–$X^2\Pi$; 434 nm, C–B) and H atoms (656 nm, H_α) were observed, as shown in Fig. 4.10c, where VG sheets were synthesized at a relatively high rate.

Another approach to produce a-C removal species is to use combined plasma sources. The so-called 'radical injection' technique was thus proposed with coupling high-intensity plasma sources such as ICP or MW, capable of effectively producing a large amount of H atoms, to the CCP or VHFCCP system [11, 38]. As briefly mentioned in Sect. 3.2.2, CCP or VHFCCP is used to produce a large number of CH_x/CF_x radicals for large-area synthesis, while the role of remote ICP or MW is to provide H radicals as an individual high-intensity plasma source. An important advantage of this technique is that it provides an active method to control the composition of radicals in the reaction volume, since CH_x/CF_x and H radicals are produced by separate plasma sources, benefiting the controllable growth of VG nanosheets [35].

4.1.3 Argon

Ar was introduced into many PECVD processes mainly for the following reasons. First, Ar has a relatively high excitation potential and ionization potential (with first excitation potential at 11.55 eV and ionization potential at 15.76 eV) [39]. This means that the interaction between electrons and Ar atoms is mostly dominated by elastic collisions and the loss of electron energy through inelastic collisions (mainly vibrational excitation), which is the major energy transfer pathway in molecular gas environment, can be significantly reduced. This effect would make the electrons retain their energy, leading to an increase in the average electron temperature. Consequently, the addition of Ar to the carbon source and a-C etchant produces a plasma capable of providing electrons of elevated energy and the necessary level of ionization rate, which can enhance the PECVD process. This feature will further improve the plasma stability.

Another advantage of Ar addition is that it can enhance the formation of critical species for VG growth, i.e., C_2, H, and ionic clusters of carbon $C_nH_x^+$ ($n \geq 2$, $x = 1, 2, 3$). Goyette et al. [20] detected the presence of gas-phase C_2 and calculated its column density in an MW PECVD system with CH_4/H_2/Ar. They found that a high fraction of Ar favored the formation of C_2 as well as H atoms. Teii et al. [6] performed OES measurements on a TM-MW plasma and suggested a C_2 formation route via the direct dissociation reaction with gaseous feedstock, e.g.:

$$C_2H_2 + Ar \rightarrow C_2 + H_2 + Ar. \tag{4.1}$$

They found that a higher Ar concentration enhances the C_2 production, benefiting the high degree of graphitization on the surface at a relatively low temperature. As shown in Fig. 4.11, the VG (i.e., labeled as CNW in the figure) growth temperature decreased from 1000 to 650 °C with increasing Ar concentration from 30 to 70 % [6]. Vizireanu et al. [18] suggested that the presence of Ar ions and metastables could help the production of C_2H_2 molecular ions excited on vibrational and rotational levels and the formation of H atoms:

$$C_2H_2 + Ar^{+,m} \rightarrow C_2H_2^{+,*} + Ar, \tag{4.2}$$

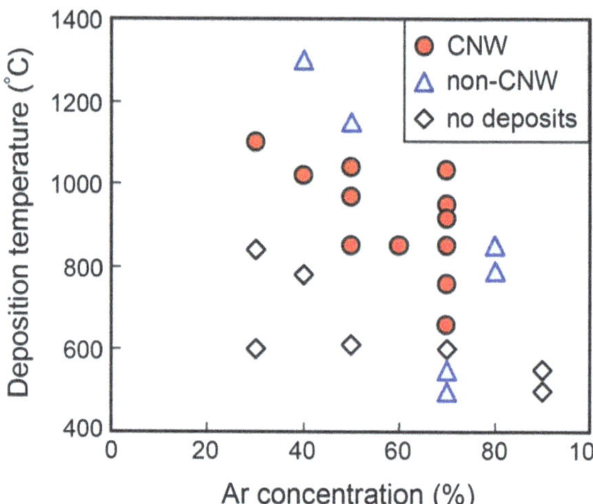

Fig. 4.11 Dependence of deposition temperature on the Ar concentration for Ar/N_2/C_2H_2 TM-MW plasma. Reprinted with permission from [6]. Copyright 2009 AIP Publishing LLC. Information on plasma type and growth conditions can be found in Table 3.1

$$H_2 + Ar^+ \rightarrow ArH^+ + H \rightarrow Ar + H_2^+. \quad (4.3)$$

The subsequent formation of higher mass clusters, which were considered as the building species for VG nanosheet growth (as aforementioned in Sect. 4.1.1), could be realized via the polymerization process involving $C_2H_2^+$ and other low-mass ions such as C_2H^+, C_2^+, CH^+, and CH_2^+, etc. [18].

Ar also offers high stability and an extremely low reaction rate, benefiting the purity of the as-grown materials. Finally, Ar is abundant and thus less expensive than other noble gases such as He.

4.1.4 Nitrogen

N_2 can work as an efficient a-C etchant in the PECVD systems for VG growth. Soin et al. [7] and Shang et al. [40] reported the synthesis of VG using a TM-MW plasma reactor with the feedstock gas of CH_4/N_2 mixture, respectively. Both studies indicated the advantage of using N_2 instead of H_2 in the removal of excessive a-C. Shang et al. [12] reported successful ultrafast growth of VG nanosheets using a CH_4/N_2 precursor in a TM-MW reactor at a temperature higher than 1000 °C, and they attributed the high growth rate of 96 μm/h partially to the stronger a-C etching ability of N_2. It has been mentioned in the literature on PECVD growth of CNTs [41] and diamond film [42] that N_2 can promote the dissociation of CH_4 to lose H atoms and to form C_2 and CN species, both of which were observed in N_2-containing PECVD systems for VG growth [6].

Meanwhile, the addition of N_2 can change the morphology and structure of the as-grown VG by substituting C atoms and introduce additional defects. Takeuchi et al. [23] conducted research on the comparison of electrical conduction between

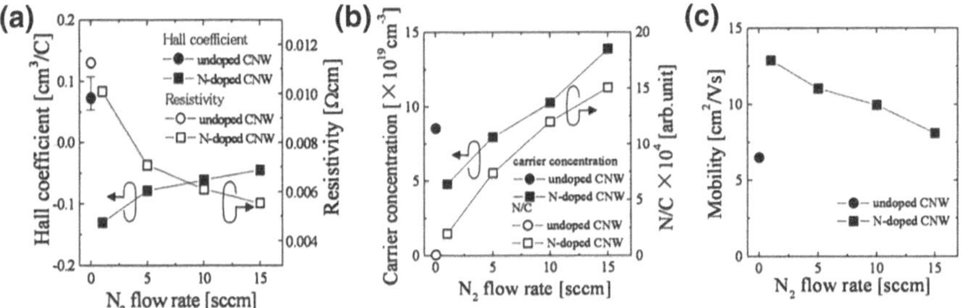

Fig. 4.12 Variations in **a** Hall coefficients and resistivity, **b** carrier concentration and N content in VG, and **c** Hall mobility as a function of N_2 flow rate during the growth process. Reprinted with permission from [23]. Copyright 2008 AIP Publishing LLC. Information on plasma type and growth conditions can be found in Table 3.1

undoped and N-doped VG samples. They found that with N_2 addition, the as-grown VG shows less alignment and highly branching morphology, but there was no obvious difference in the height and interlayer spacing.

Compared with the slight variation of morphology, the electrical conduction type presented a dramatic change with a transition from *p*-type for undoped VG to *n*-type for N-doped VG. As shown in Fig. 4.12, the resistivity and Hall coefficient of the undoped VG were obviously higher than those of the N-doped VG; the carrier (electron) concentration increased with an increase of N_2 concentration for the *n*-type N-doped VG; the charge mobility of the N-doped VG was higher than that of the undoped VG, but decreased with a further increase in the N_2 concentration. This observation was further confirmed by Teii et al.'s [6] Hall- and Seebeck-effect measurements, and thus provides an alternative for electrical conduction control on the as-grown VG, providing the potential for functional device applications. The manipulation of transport properties in VG can pave the way for applications in chemical sensing, transistors, and conductive thin films.

4.1.5 Gas Proportion

The gas proportion plays a significant role in the VG synthesis. It not only is capable of, to a certain extent, determining the type of final deposits (e.g., a-C, CNTs, diamond, and/or VG), but also strongly influences the morphology and structure of the as-grown VG.

Wu et al. [43] reported a series of deposits obtained from the same CH_4/H_2 TE-MW system with different gas ratios. As shown in Fig. 4.13a–f, carbon structures of a-C, VG sheets, CNTs/carbon nanofibers, 2-D nanographite sheets, and columnar structure of a-C were obtained with an increasing H_2 concentration. Further increase of an H_2 concentration to a rather high value will lead to the formation of diamonds. Wang et al. [14] studied the influence of a CH_4 concentration in a CH_4/H_2

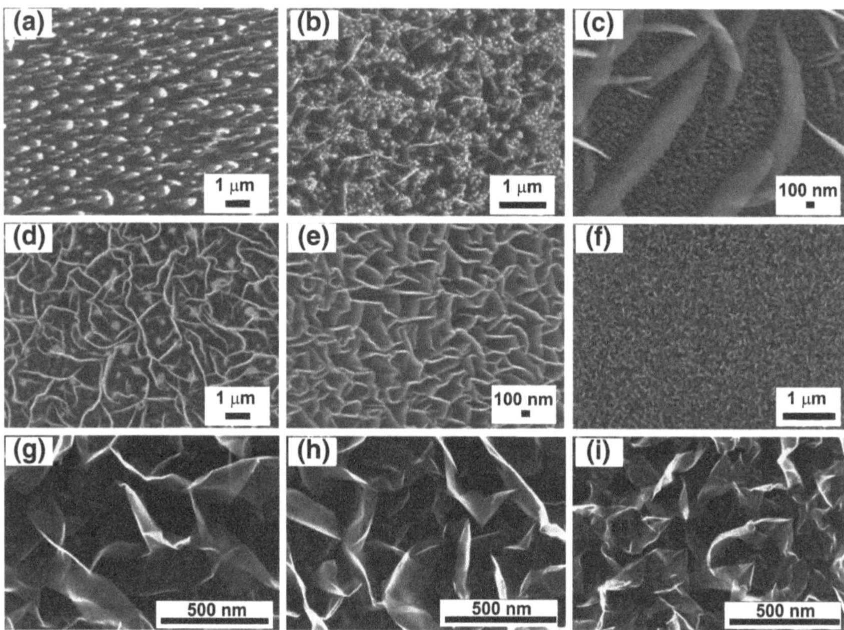

Fig. 4.13 SEM images of the deposits obtained from a TE-MW system at different H_2/CH_4 flow rate ratios: **a** 30, **b** 15, **c** 10, **d** 6, **e** 4, **f** 1. Scale bars: **a**, **b**, **d**, and **f** 1 μm; **c** and **e** 100 nm. Reprinted with permission from [43]. Copyright 2004 The Royal Society of Chemistry. SEM images of VG grown with an ICP reactor at different CH_4 concentrations on Si substrates: **g** 10 % CH_4; **h** 40 % CH_4; **i** 100 % CH_4 in an ICP system. Reprinted with permission from [14]. Copyright 2004 Elsevier. RF power: 900 W, deposition temperature: 680 °C, pressure: 12 Pa, growth time: 20 min. Information on plasma type and growth conditions can be found in Table 3.1

mixture on the VG growth performance in an ICP system. A high CH_4 concentration leads to the high nucleation tendency and consequently decreases the sheet lateral size and meanwhile increases the VG density, as shown in Fig. 4.13g–i. It is hard to define a certain optimum feedstock gas ratio for all the PECVD systems, even for the same type, due to the difference in plasma characteristics and operating conditions. Generally, for CH_4/H_2 PECVD systems, a gas ratio of CH_4:H_2 larger than 1:20 will be suitable for the growth of VG, rather than the formation of CNTs, diamond, or other nanostructures.

The proportion of the a-C etchant source in the precursor is a very important parameter for the VG growth, where its dual effects on the VG growth should be well balanced. Taking the CH_4/H_2 system as an example, H atoms etch a-C, sp^2, and sp^3 hybridized carbon simultaneously with different etching rates, which means they promote the formation of nanoislands for initial nucleation and meanwhile bombard with the as-grown VG sheets. Consequently, the number of layers in the as-grown VG sheet network decreased from 8–10 to 2–4, as reported by Zhang et al. [8] for their CH_4/H_2 TE-MW system. A similar observation was obtained when H_2O was used as the etchant. We studied the influence of H_2O

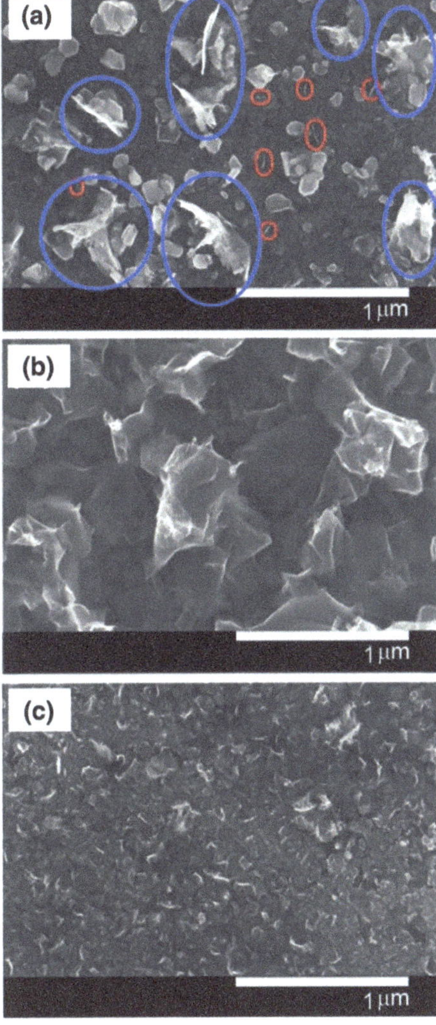

Fig. 4.14 SEM images of the deposits formed in the discharges with different H_2O addition (**a** relative humidity = 18.9 %; **b** relative humidity = 38.8 %; **c** relative humidity = 62.1 %). *Red circles* in (**a**): nanoislands; *blue circles* in (**a**): large size nanosheets. Reprinted with permission from [26]. Copyright 2011 Elsevier. Information on plasma type and growth conditions can be found in Table 3.1

content on the morphology and structure of the VG grown by an atmospheric normal glow discharge [26]. For a relatively low humidity value of 18.9 %, as shown in Fig. 4.14a, very few nanoislands (highlighted by red circles) were formed, indicating poor initial nucleation of VG; however, the lateral size of some VG layers (highlighted by blue circles) could reach ~500 nm. On the contrary, as shown in Fig. 4.14c, a relatively high relative humidity value of 62.1 % led to the formation of nanoislands of a high density, while very few large VG layers were observed. If the concentration of OH radicals reaches a certain high level, it will also prevent the subsequent formation of graphitic fragments and VG nanosheets, most likely due to undesirable oxidation erosion. An overly high relative humidity value (such as ~80 %) could significantly degrade the stability of the glow discharge.

An optimum relative humidity of ~40 % was reported for the system with the best morphology of VG layers, as shown in Fig. 4.14b.

4.2 Temperature

Thermal and electrical energies are introduced into the PECVD system in the form of heat and plasma, respectively. VG growth by a PECVD process usually can be operated at a relatively lower temperature compared with those processes using thermal CVD (T-CVD), due to the presence of energetic electrons and active species in the plasma. With good control on the plasma chemistry, the growth temperature could be lowered. As mentioned in Sect. 4.1, VG growth can be realized at a lower temperature with a higher Ar flow rate [6].

Most PECVD VG growth processes used external resistive heating on the substrate, except for MW PECVD systems where the microwave leads to substrate heating. The interaction between low-temperature plasmas and the solid substrate surface includes both the thermal and the momentum transfer [44], which means the plasmas, especially high-intensity ones, will not only provide heat for the substrate but also possibly lead to an ion etching on the substrate surface. Krivchenko et al. [45] carried out a seeding process on the substrate in an MW/RF CVD device for the subsequent VG growth using dc glow PECVD. They found that the treated surface was ion etched and led to the formation of active surface defects. For atmospheric dc negative normal glow discharges, not only the substrate but also the precursors needed to be heated, and thus, a furnace was used [26]. Generally, the amount of energy input for dc glow discharge is limited to avoid the glow to arc transition [46]. The low input energy level could lead to lower electron density and electron temperature. On the other hand, microwave or RF systems are capable of generating plasmas at higher power and temperature levels. Consequently, compared with MW or RF plasmas, the dc glow discharge usually relies on external heating to achieve similar effects as other high-energy plasma systems.

Substrate temperature is very important to the PECVD process since it strongly affects the surface reaction kinetics [47]. It is believed that a higher substrate temperature offers more species for nucleation and thus favors VG growth rate [31]. Generally, the substrate should be heated to at least 500–650 °C for successful VG growth in PECVD systems. Rao et al. [37] reported the growth of VG nanosheets by a CH_4/Ar ICP system at an extremely low substrate temperature of 400 °C; however, the formation of some spherical nanoparticles indicated that the low temperature has a negative influence on the VG growth. Wang et al. [14] investigated the influence of substrate temperature on the as-grown VG morphology in the CH_4/H_2 ICP system. It was reported that VG was not obtained if the substrate temperature was less than 600 °C. When increasing the substrate temperature to 630 °C, VG was produced but with a low density on the substrate surface and a relatively low growth rate, as shown in Fig. 4.15a. In contrast, VG grown at 730 °C showed a higher density, but the sheet surface was much smaller and less smooth, as shown in Fig. 4.15b. Further increase of temperature to 830 °C led to

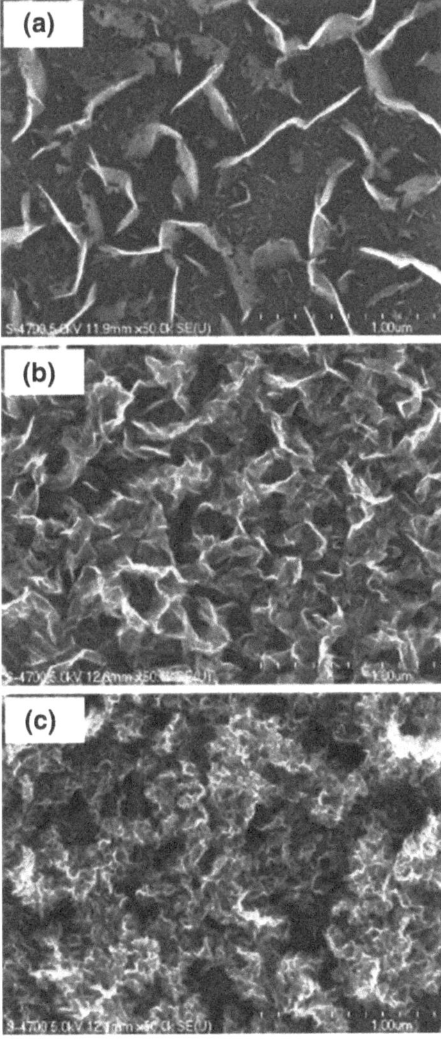

Fig. 4.15 SEM images of carbon nanosheets grown at different substrate temperatures on Si substrates: **a** 630 °C; **b** 730 °C; **c** 830 °C in an CH_4/H_2 ICP system. RF power: 900 W, CH_4 concentration: 40 %, pressure: 12 Pa, growth time: 20 min. Reprinted with permission from [14]. Copyright 2004 Elsevier. Information on plasma type and growth conditions can be found in Table 3.1

a high degree of corrugation, as shown in Fig. 4.15c. These observations can be attributed to the increase of nucleation sites on both the substrate and the as-grown layers at a higher temperature.

4.3 Summary

After discussing in the previous chapter the role of plasma source during VG synthesis in a PECVD reactor, we have introduced in this chapter how some operating parameters can affect the VG growth and the product morphology and structure.

The composition and the proportion of feedstock gases can largely affect plasma property, synthesis process, and thus the morphology and structure of VG sheets. Since a PECVD system provides both thermal and electrical energies (from heat and plasma, respectively), it is generally capable of producing VG at a relatively lower temperature than a T-CVD process that has only thermal energy. The temperature of a substrate is also a critical factor for VG synthesis in the PECVD process, since the substrate temperature can greatly influence the surface reaction kinetics.

References

1. Jain, H. G., Karacuban, H., Krix, D., Becker, H.-W., Nienhaus, H., & Buck, V. (2011). Carbon nanowalls deposited by inductively coupled plasma enhanced chemical vapor deposition using aluminum acetylacetonate as precursor. *Carbon, 49*(15), 4987–4995.
2. Bo, Z., Yang, Y., Yu, K., Chen, J., Yan, J., & Cen, K. (2013). Plasma-enhanced chemical vapor deposition synthesis of vertically-oriented graphene nanosheets. *Nanoscale, 5*(12), 5180–5204.
3. Mori, T., Hiramatsu, M., Yamakawa, K., Takeda, K., & Hori, M. (2008). Fabrication of carbon nanowalls using electron beam excited plasma-enhanced chemical vapor deposition. *Diamond and Related Materials, 17*(7–10), 1513–1517.
4. Sato, G., Morio, T., Kato, T., & Hatakeyama, R. (2006). Fast growth of carbon nanowalls from pure methane using helicon plasma-enhanced chemical vapor deposition. *Japanese Journal of Applied Physics Part 1-Regular Papers Brief Communications and Review Papers, 45*(6A), 5210–5212.
5. Chuang, A. T. H., Boskovic, B. O., & Robertson, J. (2006). Freestanding carbon nanowalls by microwave plasma-enhanced chemical vapour deposition. *Diamond and Related Materials, 15*(4–8), 1103–1106.
6. Teii, K., Shimada, S., Nakashima, M., & Chuang, A. T. H. (2009). Synthesis and electrical characterization of n-type carbon nanowalls. *Journal of Applied Physics, 106*(8), 084303.
7. Soin, N., Roy, S. S., Lim, T. H., & McLaughlin, J. A. D. (2011). Microstructural and electrochemical properties of vertically aligned few layered graphene (FLG) nanoflakes and their application in methanol oxidation. *Materials Chemistry and Physics, 129*(3), 1051–1057.
8. Zhang, Y., Du, J. L., Tang, S., Liu, P., Deng, S. Z., Chen, J., et al. (2012). Optimize the field emission character of a vertical few-layer graphene sheet by manipulating the morphology. *Nanotechnology, 23*(1), 015202.
9. Mori, S., Ueno, T., & Suzuki, M. (2011). Synthesis of carbon nanowalls by plasma-enhanced chemical vapor deposition in a CO/H-2 microwave discharge system. *Diamond and Related Materials, 20*(8), 1129–1132.
10. Chatei, H., Belmahi, M., Assouar, M. B., Le Brizoual, L., Bourson, P., & Bougdira, J. (2006). Growth and characterisation of carbon nanostructures obtained by MPACVD system using CH_4/CO_2 gas mixture. *Diamond and Related Materials, 15*(4–8), 1041–1046.
11. Shiji, K., Hiramatsu, M., Enomoto, A., Nakamura, N., Amano, H., & Hori, M. (2005). Vertical growth of carbon nanowalls using rf plasma-enhanced chemical vapor deposition. *Diamond and Related Materials, 14*(3–7), 831–834.
12. Shang, N. G., Papakonstantinou, P., McMullan, M., Chu, M., Stamboulis, A., Potenza, A., et al. (2008). Catalyst-free efficient growth, orientation and biosensing properties of multilayer graphene nanoflake films with sharp edge planes. *Advanced Functional Materials, 18*(21), 3506–3514.
13. Zhu, M. Y., Outlaw, R. A., Bagge-Hansen, M., Chen, H. J., & Manos, D. M. (2011). Enhanced field emission of vertically oriented carbon nanosheets synthesized by C_2H_2/H-2 plasma enhanced CVD. *Carbon, 49*(7), 2526–2531.

14. Wang, J. J., Zhu, M. Y., Outlaw, R. A., Zhao, X., Manos, D. M., & Holloway, B. C. (2004). Synthesis of carbon nanosheets by inductively coupled radio-frequency plasma enhanced chemical vapor deposition. *Carbon, 42*(14), 2867–2872.
15. Obraztsov, A. N., Volkov, A. P., Nagovitsyn, K. S., Nishimura, K., Morisawa, K., Nakano, Y., et al. (2002). CVD growth and field emission properties of nanostructured carbon films. *Journal of Physics D-Applied Physics, 35*(4), 357–362.
16. Obraztsov, A. N., Zolotukhin, A. A., Ustinov, A. O., Volkov, A. P., Svirko, Y., & Jefimovs, K. (2003). DC discharge plasma studies for nanostructured carbon CVD. *Diamond and Related Materials, 12*(3–7), 917–920.
17. Gruen, D. M. (1999). Nanocrystalline diamond films. *Annual Review of Materials Science, 29*, 211–259.
18. Vizireanu, S., Stoica, S. D., Luculescu, C., Nistor, L. C., Mitu, B., & Dinescu, G. (2010). Plasma techniques for nanostructured carbon materials synthesis. A case study: Carbon nanowall growth by low pressure expanding RF plasma. *Plasma Sources Science and Technology, 19*(3), 034016.
19. Hiramatsu, M., Kato, K., Lau, C. H., Foord, J. S., & Hori, M. (2003). Measurement of C-2 radical density in microwave methane/hydrogen plasma used for nanocrystalline diamond film formation. *Diamond and Related Materials, 12*(3–7), 365–368.
20. Goyette, A. N., Matsuda, Y., Anderson, L. W., & Lawler, J. E. (1998). C-2 column densities in H-2/Ar/CH_4 microwave plasmas. *Journal of Vacuum Science and Technology a-Vacuum Surfaces and Films, 16*(1), 337–340.
21. Shiomi, T., Nagai, H., Kato, K., Hiramatsu, M., & Nawata, M. (2001). Detection of C-2 radicals in low-pressure inductively coupled plasma source for diamond chemical vapor deposition. *Diamond and Related Materials, 10*(3–7), 388–392.
22. Zhu, W., Inspektor, A., Badzian, A. R., McKenna, T., & Messier, R. (1990). Effects of noble-gases on diamond deposition from methane-hydrogen microwave plasmas. *Journal of Applied Physics, 68*(4), 1489–1496.
23. Takeuchi, W., Ura, M., Hiramatsu, M., Tokuda, Y., Kano, H., & Hori, M. (2008). Electrical conduction control of carbon nanowalls. *Applied Physics Letters, 92*(21), 213103.
24. Hiramatsu, M., Shiji, K., Amano, H., & Hori, M. (2004). Fabrication of vertically aligned carbon nanowalls using capacitively coupled plasma-enhanced chemical vapor deposition assisted by hydrogen radical injection. *Applied Physics Letters, 84*(23), 4708–4710.
25. Kondo, S., Hori, M., Yamakawa, K., Den, S., Kano, H., & Hiramatsu, M. (2008). Highly reliable growth process of carbon nanowalls using radical injection plasma-enhanced chemical vapor deposition. *Journal of Vacuum Science and Technology B, 26*(4), 1294–1300.
26. Bo, Z., Yu, K., Lu, G., Wang, P., Mao, S., & Chen, J. (2011). Understanding growth of carbon nanowalls at atmospheric pressure using normal glow discharge plasma-enhanced chemical vapor deposition. *Carbon, 49*(6), 1849–1858.
27. Wang, Z., Shoji, M., & Ogata, H. (2011). Carbon nanosheets by microwave plasma enhanced chemical vapor deposition in CH4-Ar system. *Applied Surface Science, 257*(21), 9082–9085.
28. Wu, Y. H., Qiao, P. W., Chong, T. C., & Shen, Z. X. (2002). Carbon nanowalls grown by microwave plasma enhanced chemical vapor deposition. *Advanced Materials, 14*(1), 64–67.
29. Malesevic, A., Vitchev, R., Schouteden, K., Volodin, A., Zhang, L., Van Tendeloo, G., et al. (2008). Synthesis of few-layer graphene via microwave plasma-enhanced chemical vapour deposition. *Nanotechnology, 19*(30), 305604.
30. Zeng, L., Lei, D., Wang, W., Liang, J., Wang, Z., Yao, N., et al. (2008). Preparation of carbon nanosheets deposited on carbon nanotubes by microwave plasma-enhanced chemical vapor deposition method. *Applied Surface Science, 254*(6), 1700–1704.
31. Zhu, M., Wang, J., Holloway, B. C., Outlaw, R. A., Zhao, X., Hou, K., et al. (2007). A mechanism for carbon nanosheet formation. *Carbon, 45*(11), 2229–2234.
32. Kurita, S., Yoshimura, A., Kawamoto, H., Uchida, T., Kojima, K., Tachibana, M., et al. (2005). Raman spectra of carbon nanowalls grown by plasma-enhanced chemical vapor deposition. *Journal of Applied Physics, 97*(10), 104320.
33. Cheng, C. Y., & Teii, K. (2012). Control of the growth regimes of nanodiamond and nanographite in microwave plasmas. *IEEE Transactions on Plasma Science, 40*(7), 1783–1788.

34. Kondo, S., Kawai, S., Takeuchi, W., Yamakawa, K., Den, S., Kano, H., et al. (2009). Initial growth process of carbon nanowalls synthesized by radical injection plasma-enhanced chemical vapor deposition. *Journal of Applied Physics, 106*(9), 094302.
35. Hori, M., Kondo, H., & Hiramatsu, M. (2011). Radical-controlled plasma processing for nanofabrication. *Journal of Physics D-Applied Physics, 44*(17), 174027.
36. Zhu, M., Wang, J., Outlaw, R. A., Hou, K., Manos, D. M., & Holloway, B. C. (2007). Synthesis of carbon nanosheets and carbon nanotubes by radio frequency plasma enhanced chemical vapor deposition. *Diamond and Related Materials, 16*(2), 196–201.
37. Rao, B. P. C., Maheswaran, R., Ramaswamy, S., Mahapatra, O., Gopalakrishanan, C., & Thiruvadigal, D. J. (2009). Low temperature growth of carbon nanostructures by radio frequency-plasma enhanced chemical vapor deposition (low temperature growth of carbon nanostructures by RF-PECVD). *Fullerenes, Nanotubes, and Carbon Nanostructures, 17*(6), 625–635.
38. Yu, K., Bo, Z., Lu, G., Mao, S., Cui, S., Zhu, Y., et al. (2011). Growth of carbon nanowalls at atmospheric pressure for one-step gas sensor fabrication. *Nanoscale Research Letters, 6*, 202.
39. Yamabe, C., Buckman, S. J., & Phelps, A. V. (1983). Measurement of free-free emission from low-energy-electron collisions with AR. *Physical Review A, 27*(3), 1345–1352.
40. Shang, N. G., Staedler, T., & Jiang, X. (2006). Radial textured carbon nanoflake spherules. *Applied Physics Letters, 89*(10), 103112.
41. Wang, E. G., Guo, Z. G., Ma, J., Zhou, M. M., Pu, Y. K., Liu, S., et al. (2003). Optical emission spectroscopy study of the influence of nitrogen on carbon nanotube growth. *Carbon, 41*(9), 1827–1831.
42. Vandevelde, T., Wu, T. D., Quaeyhaegens, C., Vlekken, J., D'Olieslaeger, M., & Stals, L. (1999). Correlation between the OES plasma composition and the diamond film properties during microwave PA-CVD with nitrogen addition. *Thin Solid Films, 340*(1–2), 159–163.
43. Wu, Y. H., Yang, B. J., Zong, B. Y., Sun, H., Shen, Z. X., & Feng, Y. P. (2004). Carbon nanowalls and related materials. *Journal of Materials Chemistry, 14*(4), 469–477.
44. Kersten, H., Deutsch, H., Steffen, H., Kroesen, G. M. W., & Hippler, R. (2001). The energy balance at substrate surfaces during plasma processing. *Vacuum, 63*(3), 385–431.
45. Krivchenko, V. A., Dvorkin, V. V., Dzbanovsky, N. N., Timofeyev, M. A., Stepanov, A. S., Rakhimov, A. T., et al. (2012). Evolution of carbon film structure during its catalyst-free growth in the plasma of direct current glow discharge. *Carbon, 50*(4), 1477–1487.
46. Fridman, A., & Kennedy, L. (2006). *Plasma physics and engineering*. New York: Taylor & Francis.
47. Krivchenko, V., Shevnin, P., Pilevsky, A., Egorov, A., Suetin, N., Sen, V., et al. (2012). Influence of the growth temperature on structural and electron field emission properties of carbon nanowall/nanotube films synthesized by catalyst-free PECVD. *Journal of Materials Chemistry, 22*(32), 16458–16464.

Chapter 5
Atmospheric PECVD Growth of Vertically-Oriented Graphene

Abstract Vertically-oriented graphene (VG) sheets have the potential for many exciting energy and environmental applications due to their unique morphology and properties. However, it is necessary to realize the large-scale synthesis of this novel material at a low cost before we can make it economically viable for its widespread applications. It is theoretically suggested and experimentally proven that the productivity of VG is dependent on pressure, and atmospheric pressure growth could improve the efficiency and lower the cost of VG production. In the first part of this chapter, we examine how the operating pressure affects the VG production in the plasma-enhanced chemical vapor deposition (PECVD). It is also important to have the capability to grow VG on different substrates so that VG can be integrated into devices/systems for specific applications. Fortunately, the successful synthesis of VG on different substrates in PECVD processes has been realized, which is introduced in the second part of this chapter.

Keywords Atmospheric pressure · Carbon nanotube · Electric field · Graphene · Plasma-enhanced chemical vapor deposition · Orientation · Pressure · Patterned growth · Specific surface area · Substrate · Vertically-oriented graphene

5.1 How Pressure Affects the Productivity of VG During PECVD

Another important parameter significantly influencing the plasma properties and growth process is the pressure [1]. As briefly mentioned in Sect. 3.2.3, the dependency of gaseous ionization over pressure is described by Paschen's Law [2]. Qualitatively, Paschen's curve predicts that the breakdown voltage decreases as the product of electrode gap *d* and gas pressure *p* decreases. However, at some critical

---

Part of this chapter was adapted from our review article "Plasma-Enhanced Chemical Vapor Deposition Synthesis of Vertically-oriented Graphene Nanosheets," *Nanoscale*, **5**(12), 5180–5204, 2013 (DOI: 10.1039/C3NR33449J)—Reproduced by permission of The Royal Society of Chemistry.

J. Chen et al., *Vertically-Oriented Graphene*, DOI 10.1007/978-3-319-15302-5_5

$p \cdot d$ value, the breakdown voltage reaches a minimum and then increases at successively smaller $p \cdot d$ values. This is because the mean free path of electrons is inversely proportional to the pressure. So at a lower pressure, the mean free path of electrons can exceed the dimension of the plasma system, and more energy is required to overcome the rapid loss of electrons to surfaces [3].

Although a minimum voltage is usually desirable for plasma treatment, most PECVD processes are conducted at a $p \cdot d$ value smaller than the critical value. For any given PECVD system, d is usually fixed, which means that the pressure has to be lower than the critical pressure value. As mentioned previously, the lower pressure helps to achieve a higher electron temperature between two successive collisions due to the increase of the mean free path and to subsequently increase the ionization rate and maintain the stability of the plasma. However, a low pressure also means a low growth rate, while a higher plasma pressure allows a larger volume of feedstock gas input, which could possibly realize the massive production of vertically-oriented graphene (VG) at a relatively high growth rate. We demonstrated the synthesis of VG at atmospheric pressure by using a negative normal glow discharge [4]. The atmospheric growth was realized due to the extremely high local electric field near the pin tip and the external heating to the feedstock gas by a furnace. Taking advantage of the high gas flow rate at an elevated pressure, the VG growth rate could reach as high as ~300 nm/min in lateral size in the first 3 min, while the thickness of the individual VG sheet (less than 10 layers of graphene) was maintained at smaller than 10 nm.

The growth rate was not simply proportional to the pressure or the gas flow rate. It is a very complex plasma chemistry process in which the pressure influences the plasma energy and the formation of active species as well. Takeuchi et al. [5] conducted research on the influence of operating pressure on the H and C atom densities in a C_2F_6/H_2 VHFCCP + MW PECVD system, as well as the morphology and structure of the as-grown VG. As shown in Fig. 5.1a, with an increase of pressure from 13.3 to 80 Pa, the H density increases while the C density almost remains a constant. We try to explain this observation as follows. In the current VHFCCP + MW hybrid PECVD system, H and C atoms were produced from the dissociation reactions of H_2 and C_2F_6 in a 250 W MW reactor and a 270 W VHFCCP reactor, respectively. For a fixed reactor volume, the as-obtained H and C densities were simultaneously determined by the electron number density, electron energy, and reactant concentration. With the increasing pressure of a reactor and a fixed volume, the concentration of the gaseous precursor increased while the electron energy decreased due to the reduction in the electron mean free path. For H atom formation in a high-intensity MW plasma, the plasma energy was sufficient to disassociate the increasing H_2 molecules even though the electron energy was lowered, which consequently led to the increase of H atom concentration. On the other hand, the C atom formation in the VHFCCP system with a relatively lower intensity was not only dominated by the increasing gaseous concentration; the decreasing limited plasma energy was incapable of dissociating the increasing C_2F_6 molecules, and thus the C atom

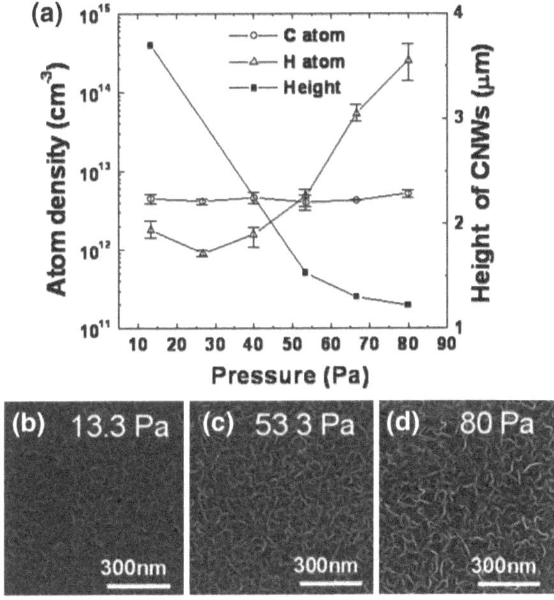

Fig. 5.1 **a** H and C atom densities in C_2F_6/H_2 VHFCCP + MW PECVD and height of the grown VG as a function of the total pressure during the formation of VG at a microwave power of 250 W and a VHF power of 270 W. **b–d** SEM images of VG grown at pressures of 13.3, 53.3, and 80.0 Pa, respectively. Reprinted with permission from [5]. Copyright 2009 AIP Publishing LLC. Information on plasma type and growth conditions can be found in Table 3.1

formation showed an almost saturated state. The increasing H atoms, induced by the increasing pressure, further influenced the morphology and structure of the as-grown VG, leading to a larger interlayer spacing but a lower VG height, as shown in Fig. 5.1b–d.

5.2 Atmospheric PECVD Growth of VG on Various Substrates

The growth of VG on various substrates is important for the integration of VG into devices/systems for different applications. In this section, we showcase some successes in producing VG at the atmospheric pressure on a range of substrates using PECVD. Initially, substrates were sputter-coated with transition metals as catalysts and the growth of VG was typically carried out in a TE-MW system [6]. Later studies demonstrated that VG growth could be realized in catalyst-free systems on arbitrary substrates that can withstand a certain heating temperature, including metals, semiconductors, and insulators (Cu [7, 8], Ni [9–12], Au [10, 13], Al [14], W [14, 15], Mo [14], Zr [14], Ti [14], Hf [14], Nb [14], Ta [14], Si [5, 12, 14, 16–29], Cr [14], stainless steel [11, 14, 15], carbon particle [30], crystal [31], GaAs [8], quartz [32], SiO_2 [10, 14, 16], Al_2O_3 (sapphire) [8, 14, 16], etc.). Until now, successful VG growth was demonstrated on convenient planar substrates, cylindrical substrates, and CNT substrates. Pattern-growth on selective areas was realized as well for possible nanoelectronic applications.

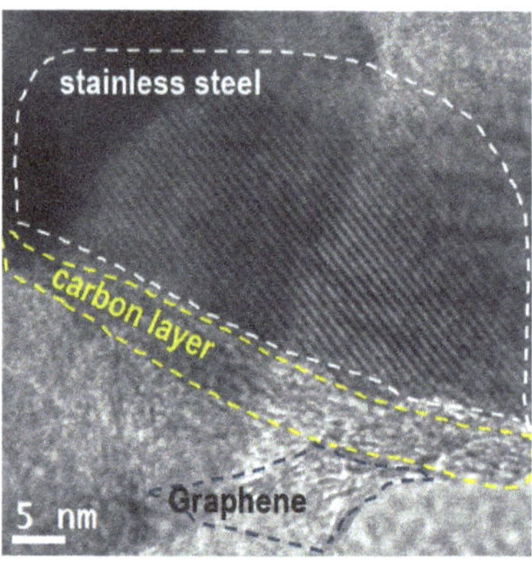

Fig. 5.2 HRTEM image of the joint between VG and stainless steel wire. Reprinted with permission from [34]. Copyright 2011 The Royal Society of Chemistry

5.2.1 Pretreatment

A routine pretreatment on the growth substrate is the ultrasonic cleaning in acetone or ethanol. Subsequent treatments could include the removal of a native oxide layer by a hydrofluoric acid solution and rinsing with deionized water [19, 28]. Teii et al. [19] suggested that VG growth was rather spontaneous and limited by the gas-phase conditions but not the surface conditions, different from that of the diamond growth. However, substrate material can affect the binding strength between the VG and the substrate; practical applications usually call for the robust growth of VG on a substrate. Generally, VG grown on carbide-forming materials could provide stronger adhesion to the substrate [33]. It has been demonstrated that a stainless steel substrate showed better VG- binding strength than a tungsten counterpart, according to Bo et al.'s [15] bath sonication treatment experiments. As shown in Fig. 5.2, VG sheets were strongly connected to the stainless steel.

Only some systems employing parallel-plate dc plasma need a particular pretreatment on the substrate to enhance VG growth. In Krivchenko et al.'s dc glow PECVD growth of a VG process, the substrate surface was pretreated by an RF/MW plasma or scratched by a diamond powder [33, 35]. The purpose of the above pretreatment was to ensure a higher nucleation density, decreased nucleation time, and improved adhesion.

5.2.2 Patterned Growth

Selective growth in a well-defined configuration could favor the construction of integrated devices and systems for nanoelectronic applications. For example, Tyler et al. demonstrated the pattern deposition of VG on the buried cathode lines of a back-gated milliampere-class field emission device by the growth of VG on the entire area, including both the background oxidized silicon wafer and the cathode lines, followed by the etching of VG at unwanted areas, to make sure that VG remained only in the centers of the cathode lines, benefiting the field emission performance [10]. Other practices of pattern-growth were commonly conducted with the pretreatment of growth substrates. Although Rao et al. [25] suggested that there was no preference for a particular substrate from the viewpoint of its growth, growth enhancement was observed on thin layers or nanoparticles of Ti, Co, Pt, and Fe [36]. Thus, depositing patterned catalyst layers on the normal substrate provides an alternative for pattern-growth of VG. For example, the VG grown on the Ti film was found to be 1.6 times higher than that on the Si substrate for the same growth conditions [36]. Wu et al. [6] also reported the pattern-growth by separating the substrate using deep trenches. Another alternative for VG pattern-growth is through controlling the electric field on the selected substrate area. We reported that patterned growth of VG nanosheets could be achieved through strategically patterning gold electrodes on the SiO_2 surface by a normal glow dc PECVD system [13]. As the first step, the growth substrate (SiO_2 wafer) was patterned as a network of squares and the letters U, W, and M, and then sputter-coated with gold electrodes of 1–5 μm wide. During the VG growth, the gold network, including the squares and letters U and M, was grounded, whereas the letter W was floating. Modeling results of the electric field at 50 nm above the substrate indicated that the electric field above the grounded gold stripe was much larger than that above the isolated gold stripe, and the electric field above the SiO_2 was the lowest, as shown in Fig. 5.3a–c. This configuration and electrical design would result in the electric field difference on different parts of the substrate and further induce non-uniform ion flux distribution and finally the difference in growth rate. As shown in Fig. 5.3d–g, with 2-min growth, VG was synthesized on the grounded square network and letters U and M, while there was no VG obtained on the isolated letter W. Further growth with a longer duration would lead to the formation of VG on the isolated letter W, but with a lower VG height. Bo et al. reported the dependence of VG morphology and structure on the local current density for the atmospheric dc normal glow discharge PECVD process [4]. Through controlling the growth time, it is possible to realize pattern-growth of VG with the designed morphology and structure on the selected area.

5.2.3 Cylindrical Substrate

Most PECVD reactors were used for the synthesis of VG sheets on planar substrates, while some practical applications call for the growth of nanostructures on non-planar (e.g., cylindrical) surfaces. Bo et al. proposed a new reactor design for continuous synthesis of VG sheets on cylindrical wire substrates using a modified atmospheric dc normal glow PECVD system [15]. Different from the previous practice where a planar substrate was fixed and mounted in a furnace, the newly designed PECVD process was conducted in a four-way quartz cross-tube outside the furnace to enable the free rotation/movement of the cylindrical growth

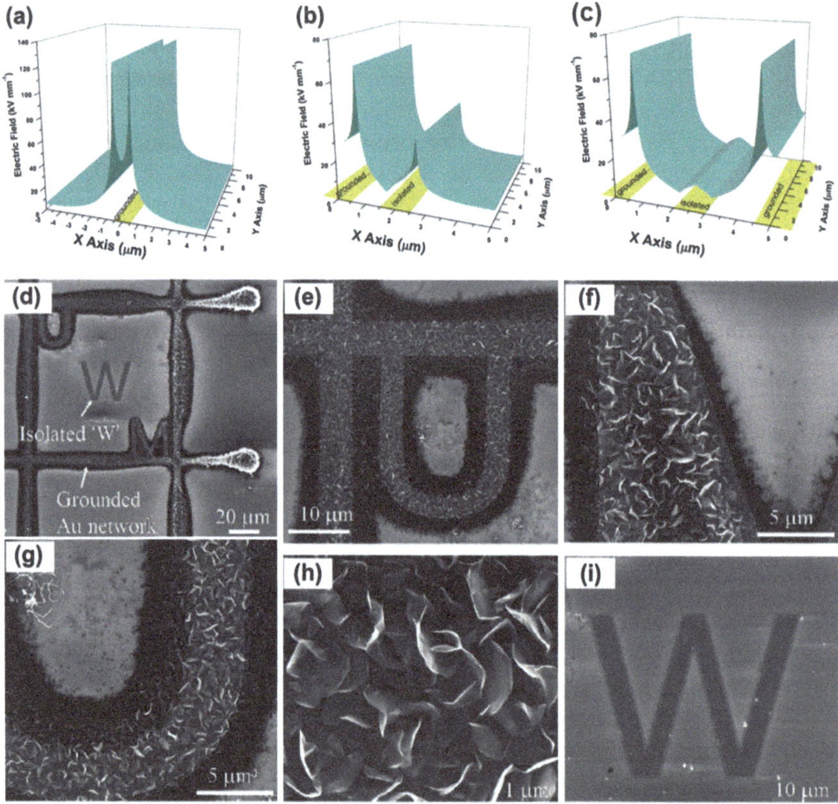

Fig. 5.3 Modeling results of the electric field at 50 nm above the growth substrate with **a** a single-grounded gold stripe, **b** a grounded gold stripe and an isolated gold stripe, and **c** two-grounded gold stripes and an isolated gold stripe. SEM images of patterned VG: **d** an overview of the VG distribution on the patterned substrate with a grounded gold square network and the letters U and M, **e** a close view showing VG growth on the grounded network and the letter U, **f**, **g** a close view showing VG growth on letters M and U, **h** enlarged view of VG showing vertical orientation of graphene nanosheets with a lateral dimension larger than 1 μm, **i** no VG growth was found on the isolated letter W. Reprinted with permission from [13]. Copyright 2011 American Chemical Society

substrates. A U-shaped reactor was used to provide additional residence time for the feed gases inside the furnace and thus enhance the heat transfer. During the VG growth, the simultaneous rotation and axial movement of the cylindrical substrate in a controlled fashion (i.e., the so-called dynamic mode synthesis) led to the uniform growth of VG. As shown in Fig. 5.4a, after 3-min synthesis, the length of the VG growth region could reach ~30 mm, including a uniform growth region of about 20 mm. Moreover, in the uniform growth region, VG morphology showed excellent uniformity in both the circumferential and the axial directions, as shown in Fig. 5.4b–d, which are the SEM images of the deposits at three different

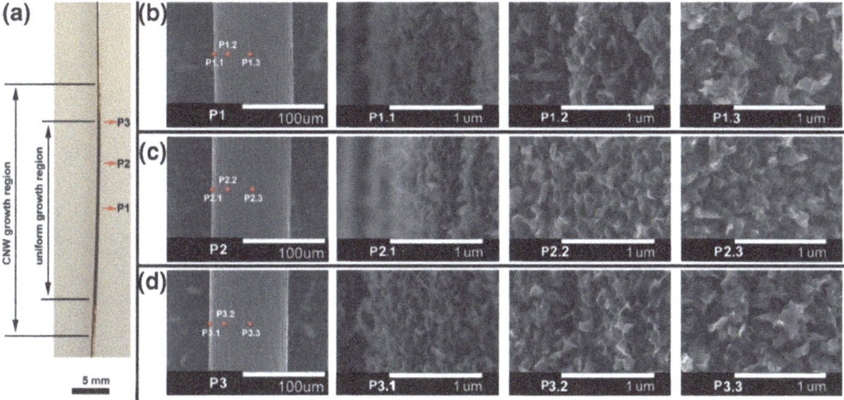

Fig. 5.4 **a** Optical image of the stainless steel wire after 3-min dynamic mode VG synthesis in a modified dc normal glow PECVD system; **b–d** SEM images of the deposits at three locations in the uniform growth region. Reprinted with permission from [15]. Copyright 2011 AIP Publishing LLC

locations on the wire surface. The above VG growth on a cylindrical substrate could benefit the immediate application of VG-coated metallic wires, such as the nanosized corona discharge plasma for indoor electrostatic devices [34].

5.2.4 CNT Substrate

Compared with those grown on planar substrates, VG grown on CNTs could lead to the full use of all three of its dimensions. VG-on-CNT hierarchical nanostructure has attracted certain interest since it provides a higher surface area due to the exposed graphene edges, compared with the pristine CNTs, and thus could possibly benefit applications such as electrochemical electrodes, catalyst supports, and gas sensors [37, 38]. Two methods of producing VG-on-CNT hybrid nanostructure will be described as follows.

Parker et al. reported the synthesis of graphenated carbon nanotubes (g-CNT) in a 915 MHz MW PECVD reactor with 10 kW maximum input power [38, 39]. In this system, vertically aligned CNT array was first produced and then further in situ growth led to the formation of small foliates from the CNT sidewall. With well-controlled growth conditions, g-CNTs with different VG densities were obtained, as shown in Fig. 5.5. High-density VG sheets were synthesized on CNTs, as shown in Fig. 5.5c, via the secondary nucleation on the initially grown foliates. A possible growth mechanism based on the buckling of CNTs with a longer growth duration was proposed. This method of growing g-CNTs required a relatively large diameter of CNTs to ensure an easy buckling, benefiting the formation of foliate nucleation sites, and strain relaxation of external CNT walls; the smallest diameter CNT exhibiting graphitic foliates in this work was reported as 80 nm.

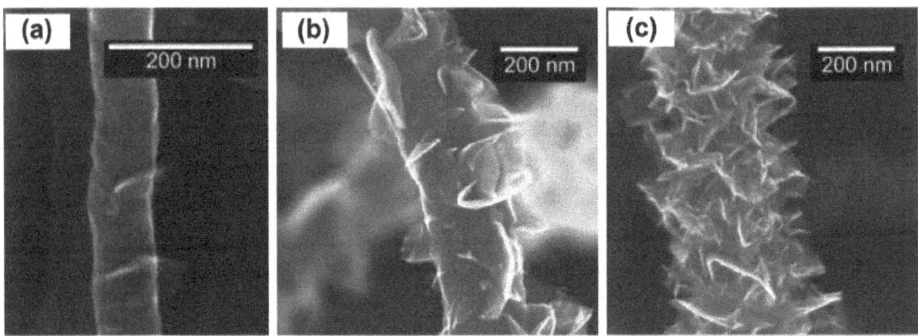

Fig. 5.5 SEM images of graphenated carbon nanotubes (g-CNTs): **a** low-density graphene foliates on a CNT, **b** medium-density graphene foliates on a CNT, **c** high-density graphene foliates on a CNT. Reproduced with permission from [38]. Copyright 2012 Cambridge University Press

We conducted VG growth on both the horizontally dispersed and vertically aligned CNTs by an atmospheric dc normal glow discharge PECVD system [37]. For this method, there was almost no requirement on the CNT diameter, where VG nanosheets were successfully grown on CNTs of ~15 nm in diameter. With a 1-min PECVD process, CNT–graphene hybrid structures were obtained with graphene sheets (<10 nm thick) spreading off the host CNTs (Fig. 5.6a, b), in contrast to the smooth surface of pristine CNTs before the graphene growth. The lateral dimension of the graphene typically ranged from 200 to 500 nm, which is consistent with a ~300 nm/min growth rate previously observed with the same technique and could be controlled by the growth time and the CH_4 concentration [4]. By simply using a different substrate, graphene growth was also performed on vertically aligned CNTs. Graphene sheets could be grown on the entire external surface of every single vertical CNT (Fig. 5.6c). The inset of Fig. 5.6c shows that although the CNT stem was uniform in diameter, a much larger VG corolla capped the tube tip, suggesting that the VG nanosheets grow faster on the tip portion of a vertical CNT where the local electric field is higher, leading to faster graphene growth near the CNT tip. Large graphene sheets with lateral dimensions of nearly 1 μm can be grown in some cases (Fig. 5.6d). The CNT/graphene interface has a significant influence on the characteristics of the CNT–graphene hybrid structure. Figure 5.6e is a representative HRTEM image that shows a few-layer graphene sheet growing directly off a host CNT; individual graphene layers can be discerned. The graphene and the host CNT are inherently bonded via sp^2 carbons, evidenced by the matching lattice fringes of the graphene and those of CNTs. A noteworthy increase of the I(2D)/I(G) ratio in Raman spectra suggested the presence of more sp^2 carbon domains after the graphene growth on CNTs (Fig. 5.6f). Compared with the method proposed by Parker et al. [38], there was no certain requirement on the CNT diameter. Fusing VG with CNT into a single unified structure represents a giant step toward engineering interfaces between constituent components in hybrid nanostructures.

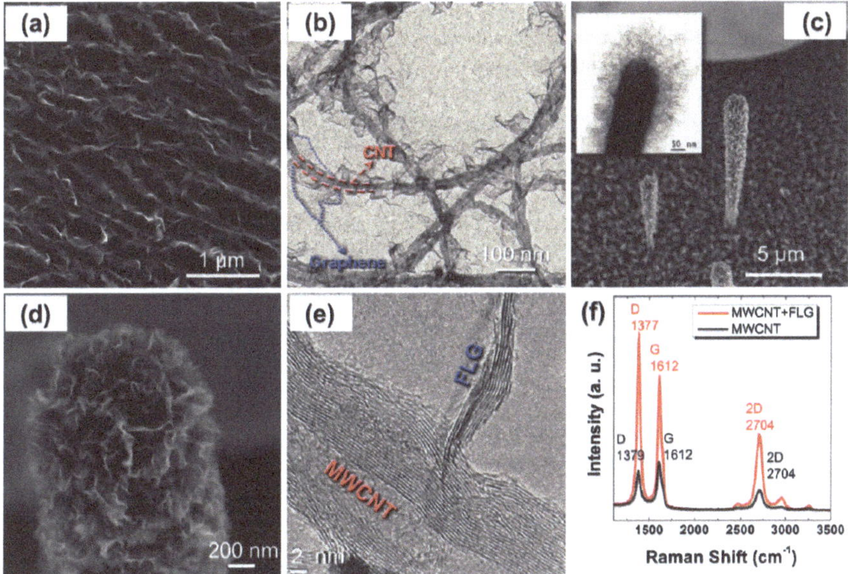

Fig. 5.6 **a** SEM image of VG-on-CNT hybrid structures. **b** TEM image of VG grown on CNTs suspended on a Cu TEM grid. **c** SEM image of VG grown on vertically aligned CNTs; the inset is a TEM image of the tip of a VG-on-CNTs, where the CNT stem (*dark*) is clearly distinguished from VG (*bright*). **d** A close view of the VG-on-CNTs tip in panel **c**. **e** HRTEM image of a VG-on-CNTs structure showing that the VG is inherently bonded to the host CNT. **f** Raman spectra of pristine CNT (*black*) and VG-on-CNTs (*red*) films. Reprinted with permission from [37]. Copyright 2011 American Chemical Society

5.3 Summary

The large-scale production of VG at a low cost is critical for the widespread use of VG in various applications. In this chapter, we have analyzed how the pressure in a PECVD reactor can affect the VG synthesis in terms of VG growth efficiency. The operation of the PECVD process at the atmospheric pressure can potentially improve the production efficiency of VG and considerably lower the cost by avoiding an expensive vacuum apparatus. The capability to grow VG on various substrates is important for the incorporation of VG into devices/systems for different applications. We have demonstrated some successes in synthesizing VG at the atmospheric pressure on a range of substrates in PECVD processes.

References

1. Bo, Z., Yang, Y., Yu, K., Chen, J., Yan, J., & Cen, K. (2013). Plasma-enhanced chemical vapor deposition synthesis of vertically-oriented graphene nanosheets," *Nanoscale, 5*(12), 5180–5204.

2. Paschen, F. (1889). Ueber die zum Funkenübergang in Luft, Wasserstoff und Kohlensäure bei verschiedenen Drucken erforderliche Potentialdifferenz. *Annalen der Physik, 273*(5), 69–96.
3. Lieberman, M. A., & Lichtenberg, A. J. (2005). *Principles of plasma discharges and materials processing* (2nd ed., pp. 1–757). New Jersey: Wiley.
4. Bo, Z., Yu, K., Lu, G., Wang, P., Mao, S., & Chen, J. (2011). Understanding growth of carbon nanowalls at atmospheric pressure using normal glow discharge plasma-enhanced chemical vapor deposition. *Carbon, 49*(6), 1849–1858.
5. Takeuchi, W., Sasaki, H., Kato, S., Takashima, S., Hiramatsu, M., & Hori, M. (2009). Development of measurement technique for carbon atoms employing vacuum ultraviolet absorption spectroscopy with a microdischarge hollow-cathode lamp and its application to diagnostics of nanographene sheet material formation plasmas. *Journal of Applied Physics, 105*(11), 113305.
6. Wu, Y. H., Yang, B. J., Han, G. C., Zong, B. Y., Ni, H. Q., Luo, P., et al. (2002). Fabrication of a class of nanostructured materials using carbon nanowalls as the templates. *Advanced Functional Materials, 12*(8), 489–494.
7. Yu, K., Bo, Z., Lu, G., Mao, S., Cui, S., Zhu, Y., et al. (2011). Growth of carbon nanowalls at atmospheric pressure for one-step gas sensor fabrication. *Nanoscale Research Letters, 6*, 202.
8. Wu, Y. H., Qiao, P. W., Chong, T. C., & Shen, Z. X. (2002). Carbon nanowalls grown by microwave plasma enhanced chemical vapor deposition. *Advanced Materials, 14*(1), 64–67.
9. Sato, G., Morio, T., Kato, T., & Hatakeyama, R. (2006). Fast growth of carbon nanowalls from pure methane using helicon plasma-enhanced chemical vapor deposition. *Japanese Journal of Applied Physics, 45*(6A), 5210–5212.
10. Tyler, T., Shenderova, O., Ray, M., Dalton, J., Wang, J., Outlaw, R., et al. (2006). Back-gated milliampere-class field emission device based on carbon nanosheets. *Journal of Vacuum Science and Technology B, 24*(5), 2295–2301.
11. Bo, Z., Wen, Z., Kim, H., Lu, G., Yu, K., & Chen, J. (2012). One-step fabrication and capacitive behavior of electrochemical double layer capacitor electrodes using vertically-oriented graphene directly grown on metal. *Carbon, 50*(12), 4379–4387.
12. Obraztsov, A. N., Volkov, A. P., Nagovitsyn, K. S., Nishimura, K., Morisawa, K., Nakano, Y., & Hiraki, A. (2002). CVD growth and field emission properties of nanostructured carbon films. *Journal of Physics D-Applied Physics, 35*(4), 357–362.
13. Yu, K., Wang, P., Lu, G., Chen, K.-H., Bo, Z., & Chen, J. (2011). Patterning vertically-oriented graphene sheets for nanodevice applications. *Journal of Physical Chemistry Letters, 2*(6), 537–542.
14. Wang, J. J., Zhu, M. Y., Outlaw, R. A., Zhao, X., Manos, D. M., Holloway, B. C., & Mammana, V. P. (2004). Free-standing subnanometer graphite sheets. *Applied Physics Letters, 85*(7), 1265–1267.
15. Bo, Z., Cui, S., Yu, K., Lu, G., Mao, S., & Chen, J. (2011). Note: continuous synthesis of uniform vertical graphene on cylindrical surfaces. *Review of Scientific Instruments, 82*(8), 086116.
16. Hiramatsu, M., Shiji, K., Amano, H., & Hori, M. (2004). Fabrication of vertically aligned carbon nanowalls using capacitively coupled plasma-enhanced chemical vapor deposition assisted by hydrogen radical injection. *Applied Physics Letters, 84*(23), 4708–4710.
17. Chuang, A. T. H., Boskovic, B. O., & Robertson, J. (2006). Freestanding carbon nanowalls by microwave plasma-enhanced chemical vapour deposition. *Diamond and Related Materials, 15*(4–8), 1103–1106.
18. Kondo, S., Hori, M., Yamakawa, K., Den, S., Kano, H., & Hiramatsu, M. (2008). Highly reliable growth process of carbon nanowalls using radical injection plasma-enhanced chemical vapor deposition. *Journal of Vacuum Science and Technology B, 26*(4), 1294–1300.
19. Teii, K., Shimada, S., Nakashima, M., & Chuang, A. T. H. (2009). Synthesis and electrical characterization of n-type carbon nanowalls. *Journal of Applied Physics, 106*(8), 084303.
20. Wang, Z., Shoji, M., & Ogata, H. (2011). Carbon nanosheets by microwave plasma enhanced chemical vapor deposition in CH4-Ar system. *Applied Surface Science, 257*(21), 9082–9085.
21. Zhu, M. Y., Outlaw, R. A., Bagge-Hansen, M., Chen, H. J., & Manos, D. M. (2011). Enhanced field emission of vertically-oriented carbon nanosheets synthesized by C2H2/H-2 plasma enhanced CVD. *Carbon, 49*(7), 2526–2531.

22. Shang, N. G., Au, F. C. K., Meng, X. M., Lee, C. S., Bello, I., & Lee, S. T. (2002). Uniform carbon nanoflake films and their field emissions. *Chemical Physics Letters, 358*(3–4), 187–191.
23. Zhang, Y., Du, J. L., Tang, S., Liu, P., Deng, S. Z., Chen, J., & Xu, N. S. (2012). Optimize the field emission character of a vertical few-layer graphene sheet by manipulating the morphology. *Nanotechnology, 23*(1), 015202.
24. Shiji, K., Hiramatsu, M., Enomoto, A., Nakamura, N., Amano, H., & Hori, M. (2005). Vertical growth of carbon nanowalls using rf plasma-enhanced chemical vapor deposition. *Diamond and Related Materials, 14*(3–7), 831–834.
25. Rao, B. P. C., Maheswaran, R., Ramaswamy, S., Mahapatra, O., Gopalakrishanan, C., & Thiruvadigal, D. J. (2009). Low temperature growth of carbon nanostructures by radio frequency-plasma enhanced chemical vapor deposition (low temperature growth of carbon nanostructures by RF-PECVD). *Fullerenes, Nanotubes, and Carbon Nanostructures, 17*(6), 625–635.
26. Kondo, S., Kawai, S., Takeuchi, W., Yamakawa, K., Den, S., Kano, H., et al. (2009). Initial growth process of carbon nanowalls synthesized by radical injection plasma-enhanced chemical vapor deposition. *Journal of Applied Physics, 106*(9), 094302.
27. Soin, N., Roy, S. S., O'Kane, C., McLaughlin, J. A. D., Lim, T. H., & Hetherington, C. J. D. (2011). Exploring the fundamental effects of deposition time on the microstructure of graphene nanoflakes by Raman scattering and X-ray diffraction. *Crystal Engineering Communication, 13*(1), 312–318.
28. Cheng, C. Y., & Teii, K. (2012). Control of the growth regimes of nanodiamond and nanographite in microwave plasmas. *IEEE Transactions on Plasma Science, 40*(7), 1783–1788.
29. Teii, K., & Ikeda, T. (2007). Effect of enhanced C-2 growth chemistry on nanodiamond film deposition. *Applied Physics Letters, 90*(11), 111504.
30. Tiwari, J. N., Tiwari, R. N., Singh, G., & Lin, K. L. (2011). Direct synthesis of vertically interconnected 3-D graphitic nanosheets on hemispherical carbon particles by microwave plasma CVD. *Plasmonics, 6*(1), 67–73.
31. Mori, S., Ueno, T., & Suzuki, M. (2011). Synthesis of carbon nanowalls by plasma-enhanced chemical vapor deposition in a CO/H-2 microwave discharge system. *Diamond and Related Materials, 20*(8), 1129–1132.
32. Takeuchi, W., Ura, M., Hiramatsu, M., Tokuda, Y., Kano, H., & Hori, M. (2008). Electrical conduction control of carbon nanowalls. *Applied Physics Letters, 92*(21), 213103.
33. Krivchenko, V. A., Dvorkin, V. V., Dzbanovsky, N. N., Timofeyev, M. A., Stepanov, A. S., Rakhimov, A. T., et al. (2012). Evolution of carbon film structure during its catalyst-free growth in the plasma of direct current glow discharge. *Carbon, 50*(4), 1477–1487.
34. Bo, Z., Yu, K., Lu, G., Cui, S., Mao, S., & Chen, J. (2011). Vertically-oriented graphene sheets grown on metallic wires for greener corona discharges: Lower power consumption and minimized ozone emission. *Energy & Environmental Science, 4*(7), 2525–2528.
35. Krivchenko, V. A., Pilevsky, A. A., Rakhimov, A. T., Seleznev, B. V., Suetin, N. V., Timofeyev, M. A., et al. (2010). Nanocrystalline graphite: Promising material for high current field emission cathodes. *Journal of Applied Physics, 107*(1), 014315.
36. Hiramatsu, M., & Hori, M. (2010). *Carbon nanowalls: Synthehesis and emerging application*. New York: Springer.
37. Yu, K., Lu, G., Bo, Z., Mao, S., & Chen, J. (2011). Carbon nanotube with chemically bonded graphene leaves for electronic and optoelectronic applications. *Journal of Physical Chemistry Letters, 2*(13), 1556–1562.
38. Parker, C. B., Raut, A. S., Brown, B., Stoner, B. R., & Glass, J. T. (2012). Three-dimensional arrays of graphenated carbon nanotubes. *Journal of Materials Research, 27*(7), 1046–1053.
39. Cui, H., Zhou, O., & Stoner, B. R. (2000). Deposition of aligned bamboo-like carbon nanotubes via microwave plasma enhanced chemical vapor deposition. *Journal of Applied Physics, 88*(10), 6072–6074.

Chapter 6
Vertically-Oriented Graphene for Sensing and Environmental Applications

Abstract As an atomic-thick layer material, graphene has a large specific surface area, a high electron mobility, and a high sensitivity to electronic perturbations from adsorption/desorption of molecules, which are attractive properties from the sensing perspective. Graphene has been widely studied for sensing applications, e.g., in biosensors and gas sensors. On the other hand, although graphene shows promise in miniaturized and high-performance sensors, many challenges still remain for the low-cost and large-scale fabrication of graphene sensors with reliable performance. Recently, vertically-oriented graphene (VG) has emerged as a new type of graphene material, and VG-based nanostructures have been explored to advance the sensing technology due to VG's unique structure and properties. Using VG as the sensing element, VG sensors have been studied for the detection of biomolecules (e.g., proteins, DNA, and bacteria) and gases (e.g., NO_2, NH_3, and H_2). The VG-based sensors have shown promising features that could be used to address some challenges of current sensing technologies for biomolecule and gas detection, e.g., high cost, low sensitivity/selectivity, and/or being unsuitable for in situ and real-time detection. Due to its vertical orientation and high conductivity, VG has also been used in green corona discharge with low ozone generation, which is another important application of VG in environmental fields. In this chapter, representative VG-based sensors and VG corona discharge are introduced and discussed.

Keywords Environmental application · Electronic biosensor · Electrochemical biosensor · Gas sensor · Corona discharge

There is a need for simple and reliable sensors suitable for trace detection in a wide spectrum of applications ranging from medical diagnosis, environmental monitoring, industrial, agricultural to lab-on-a-chip. Semiconducting nanomaterials

---

Part of this chapter was adapted from our review article "Emerging Energy and Environmental Applications of Vertically-Oriented Graphenes," *Chemical Society Reviews*, 2015 (DOI: 10.1039/C4CS00352G) — Reproduced by permission of The Royal Society of Chemistry.

J. Chen et al., *Vertically-Oriented Graphene*, DOI 10.1007/978-3-319-15302-5_6

have been widely used for biomolecule and gas sensing applications, including semiconducting metals/metal oxides [1–3] and nanocarbon-based materials [4–6]. Among those nanomaterials, CNT and graphene [7–11] are two popular sensing materials for biomolecule and gas detection due to their high sensitivity to various biomolecules and gases. The unique structures and outstanding electronic properties from CNT and graphene, e.g., small size, large specific surface area, high electron mobility, and high sensitivity to electronic perturbations from foreign molecules, offer great promise for use in miniaturized and high-performance sensors. Many studies have demonstrated their use in biomolecule and gas detections, including various target analytes. Recently, VG-based nanostructures have been explored for sensing applications, i.e., biosensors and gas sensors, due to their unique structure and properties.

6.1 VG-Based Biosensors

6.1.1 Electronic Biosensors

Depending on the specific working principle, graphene-based biosensors either use their electronic properties (e.g., high carrier mobility and high sensitivity to electronic perturbations), electrochemical properties (e.g., high catalytic activity and electron transfer rates), or unique structure (e.g., atom-layer thickness and large specific surface area) for biomolecule detection. Different from conventional graphenes, VGs draw increasing attention in biosensor applications due to their unique vertical orientation and open structure. VGs are high-performing sensing materials as their surface can be fully accessible by analytes, and the high length of exposed edges and the high in-plane carrier mobility lead to superior sensor performance.

A field-effect transistor (FET) biosensor can be fabricated by direct growth of VGs on the sensor electrodes, as shown in Fig. 6.1a, b [12]. The sensor contains patterned metal electrodes, VGs (sensing material), and gold nanoparticle (Au NP)–antibody conjugates (as a probe for analyte protein binding). The sensor configuration allows for the direct diffusion and binding of the target protein (Immunoglobulin G/IgG) to the probe protein (anti-IgG) on the Au NP. When a target protein binds to the probe protein, it causes a change in the electrical conductance of the VG nanosheet that can be measured by the external circuit/measuring system. In this sensor, the VGs showed *p*-type semiconducting characteristics in an ambient environment, which is quite similar to the characteristics of graphene or reduced graphene oxide (RGO). During the sensing test (Fig. 6.1c), with the introduction of the IgG solution to the sensor, the drain current I_d decreased and the sensor had a high sensitivity down to 2 ng ml^{-1} or 13 pM for IgG proteins.

Typically, nanomaterial-based electronic sensors have a rapid response to target analytes, which brings great potential to this type of sensors in real-time biosensing and rapid diagnostics. Figure 6.2a shows the dynamic responses of VG sensors to different IgG concentrations with and without the probe proteins, respectively. The results clearly show that the sensor conductivity decreased correspondingly

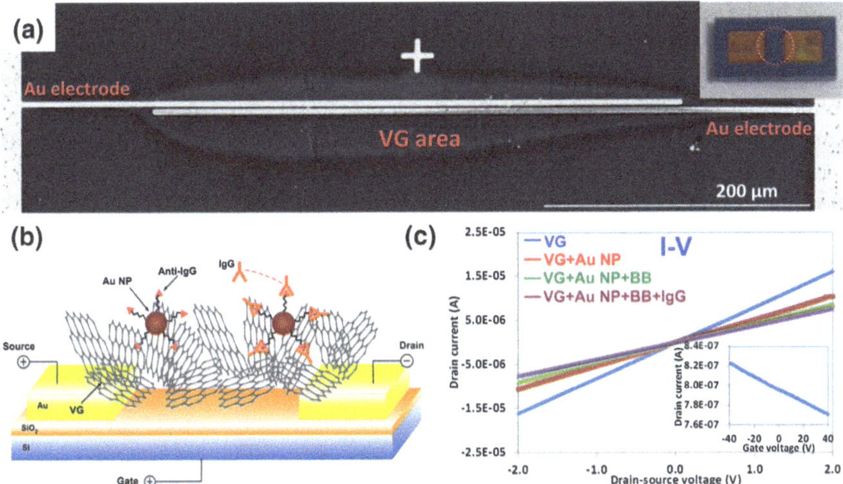

Fig. 6.1 **a** SEM image of a sensor fabricated by direct growth of VG between gold electrodes. Inset is a digital image of the sensor electrode. **b** Schematic of the VG electronic sensor. The binding of the analyte protein (IgG) to the antibody on the VG causes a change in the electrical conductance of the sensor. **c** Sensor conductivity changes with probe antibody labeling and analyte protein detection. Inset is the FET transport characteristic of the VG sensor. Reprinted with permission from [12]. Copyright 2013 Nature Publishing Group

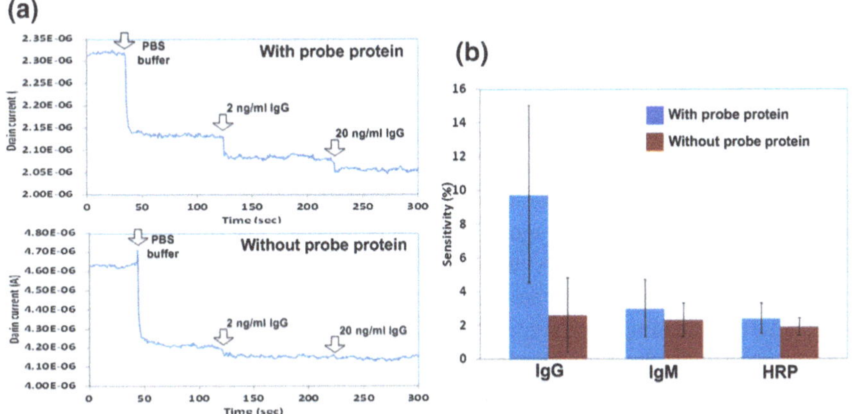

Fig. 6.2 **a** Dynamic response of the VG sensor exposed to different concentrations of IgG with (*top*) and without (*bottom*) probe proteins. **b** Comparison of the sensor sensitivity in response to complementary IgG (2 ng/mL), mismatched IgM (0.2 mg/mL), and mismatched HRP (0.2 mg/mL). Error bars were obtained from multiple samples. Reprinted with permission from [12]. Copyright 2013 Nature Publishing Group

to the injection of the proteins and the response time is on the order of seconds. The critical function of the probe was also confirmed through this study since the sensors without probes (anti-IgG) showed much lower responses compared with those of the sensors with probes. Those responses (Fig. 6.2a bottom) are believed to originate from the non-specific binding of the IgG proteins to the VG surface. The specificity/selectivity of a biosensor is another critical parameter for sensor applications. As shown in Fig. 6.2b, the sensor responses from the mismatched Immunoglobulin M (IgM) (3.0 %) and horseradish peroxidase (HRP) (2.4 %) are significantly smaller than those from the complementary IgG (9.8 %), confirming the high specificity of the sensor.

By comparing the VG sensor with conventional horizontal graphene-based electronic sensors, the vertical orientation and open structure of VG increase the accessible area of the device to analytes, thereby increasing the sensitivity of the sensor. In addition, direct PECVD growth of VGs on sensor electrodes could achieve higher stability and reproducibility than the drop-casting method, which is commonly used in graphene biosensor fabrication. The high stability and reproducibility are attributed to the stronger binding between the VG and the sensor electrode as compared with the drop-casting deposition.

6.1.2 Electrochemical Biosensors

In addition to electronic sensors based on the conductivity change in the sensing element, VG-based electrochemical sensors were also reported for biomolecule detection. The electrochemical sensors use the electrocatalytic activity of the sensing electrode in a redox system for analyte detection. Typically, the sensor records reaction fingerprints (e.g., oxidation or reduction peaks) of the analyte in the cyclic voltammetry (CV) measurements, and the intensity of current peaks can be related to the analyte concentration. Figure 6.3a shows the CV curves of a VG-based electrochemical sensor for the simultaneous detection of dopamine (DA), ascorbic acid (AA), and uric acid (UA) [13]. The three analytes registered three pairs of oxidation and reduction peaks in the CV curves at different potentials, which could be used as the fingerprints to qualitatively determine the analytes. In particular, the detection of DA with various concentrations (1–100 μM) was demonstrated in the presence of common interfering agents of AA and UA, as shown in Fig. 6.3b, indicating high sensitivity and selectivity of the sensor. By testing the sensors with different DA concentrations, the relation between the reaction current and the DA concentration was obtained, which could be used as a calibration chart to estimate the analyte concentration directly from the electrochemical measurements.

The excellent sensor performance of the electrochemical VG sensor is attributed to the enhanced electron transfer from the VG electrode. In general, high electronic density of states (DOS) in the electrode leads to the increased electron transfer in a redox system [13]. Compared with the basal planes of graphene with a low DOS at the Fermi level, defects (such as kinks, steps, and vacancies) on the edges of VGs can produce localized edge states between the conduction and the valence bands, resulting in a high DOS near the Fermi level. The investigation on

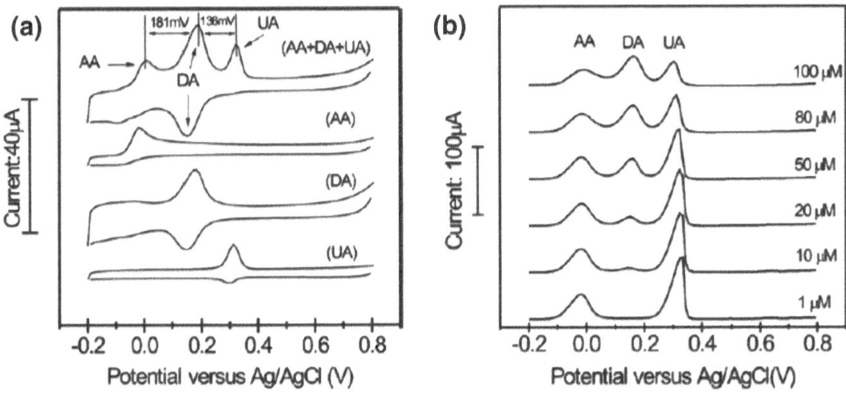

Fig. 6.3 **a** CV profiles of the VG-based electrochemical sensor in the solution of 50 mM, pH 7.0 PBS with individual 1 mM AA, 0.1 mM DA, and 0.1 mM UA, and their mixtures. Each analyte has a characteristic pair of redox peaks in the CV, and the mixture shows distinct peaks for each analyte. **b** Differential pulse voltammetric profiles of the VG electrode with 1 mM AA, 0.1 mM UA, and different concentrations of DA from 1 to 100 μM. The DA peak current increases with the increasing DA concentration. Reprinted with permission from [13]. Copyright 2008 John Wiley and Sons

the stability of the VG electrode showed that the morphology of VGs remained unchanged after long-term cycling. This research confirms the mechanical robustness of VGs and their good electrochemical stability.

A VG-based electrochemical sensor was also reported for DNA detection. Akhavan et al. reported an ultrahigh-resolution electrochemical biosensor with reduced graphene nanowall (RGNW) for the detection of four bases of DNA (G, A, T, and C) by monitoring the oxidation signals of individual nucleotide bases [14]. The RGNW was made by depositing GO nanosheets on a graphite electrode using electrophoretic deposition followed by hydrazine reduction. Their study showed that the RGNW electrode worked perfectly for a wide concentration range (0.1 fM–10 mM) of double-stranded DNA (dsDNA) (Fig. 6.4a). The excellent performance of the RGNW electrode at such high and low concentrations was attributed to the high surface porosity and edge defects of RGNW, which inhibited the blocking of the electrode from fouling and accelerated the electron transfer between the electrode and dsDNA, respectively. The RGNW electrode also exhibited an excellent linear behavior for the current response in a wider and more sensitive range of dsDNA bases compared with the reduced graphene nanosheet (RGNS) electrode (Fig. 6.4b).

Due to the unique structure, VG sensors based on optical signals (surface-enhanced Raman spectroscopy/SERS) were also reported and showed great potential for biomolecule detection [15, 16]. In one study, the Ostrikov group studied the interaction between the VG and the gold NP-labeled goat antimouse antibody fragments (gold AB) though SERS tests. The antibody introduced significant changes in the Raman spectra of VG, which indicated that the VG platform

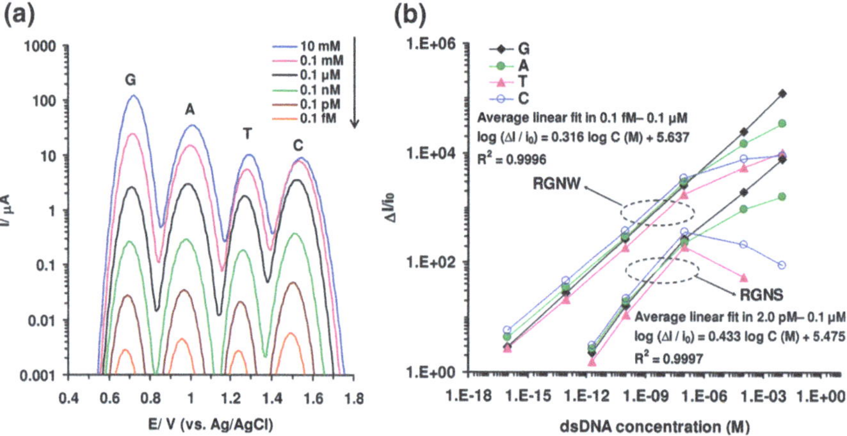

Fig. 6.4 **a** Logarithmic presentation of DPV profiles of RGNW electrodes for the detection of various concentrations of dsDNA. **b** Log–log plot of the current response of the RGNW and RGNS electrodes to the four bases (*G*, *A*, *T*, and *C*) of the dsDNA at various concentrations. Reprinted with permission from [14]. Copyright 2012 American Chemical Society

could potentially work for antigen detection with a SERS signal. Although there are limited reports on optical biosensor applications based on VG, it is believed that the vertically arranged graphene structure could be a good alternative to a flat graphene-based platform due to its unique structure and easy deposition of ordered arrays of Au NPs on their edges and basal planes.

6.2 VG-Based Gas Sensors

Air pollutants, e.g., sulfur oxides, nitrogen oxides, carbon monoxide, ammonia, volatile organic compounds, and particulates, are significant risk factors for a number of health conditions. The real-time monitoring of the air pollution is critical for the reduction of harmful effects from air pollutants to humans; this monitoring generally relies on a gas sensor to detect specific species. Graphene/RGO-based materials have been widely studied for electronic gas sensor applications due to their large specific surface areas and their high sensitivity to electronic perturbations upon gas molecule adsorption. Microsized sensors made from graphene are able to detect individual gas molecules, which change the local carrier concentration in graphene sheets. The gas-induced changes in conductivity have different magnitudes for different gases, and the sign of the change (increase or decrease in conductivity) indicates whether the gas is an electron acceptor (e.g., NO_2) or an electron donor (e.g., NH_3 and H_2). Similar to VG-based biosensors, the vertical and open structures of VGs offer large accessible surface areas for gas molecule adsorption and inhibit the agglomeration of graphene sheets during the sensor device fabrication.

The electric field distribution above the substrate guides the VG growth by PECVD and can be used for area-specific synthesis of VG-based FET gas sensors [17]. Due to the enhanced electric field, VG sheets could be selectively grown on a gold sensor electrode with different patterns. The VG-based sensor was tested with three consecutive steps that include exposure of the device to an airflow to record a base value of the sensor resistance, then to the analyte gas to register a sensing signal, and subsequently to an airflow for sensor recovery. The VG-based sensor can be operated at room temperature for NO_2 (100 ppm) and NH_3 (1 %) detection (Fig. 6.5a). The sensitivities, defined as ratios of R_{air}/R_{NO2} and R_{NH3}/R_{air} (R_{air}, R_{NO2}, and R_{NH3} are the sensor resistances in air, in NO_2, and in NH_3, respectively) of the VG sensor to NO_2 (1.57) and NH_3 (1.13), were comparable with multilayer graphene-based devices. The VGs behaved like a p-type semiconductor in an ambient environment, which was confirmed by the measurement of FET transport characteristics shown in Fig. 6.5b. The sensing mechanism is based on the adsorbed NH_3 donating electrons and neutralizing holes, which decreases the VG conductance. On the other hand, NO_2 accepts electrons from VGs, which leads to the increased VG conductance. In addition to pure VG-based gas sensors, a VG/CNT hybrid was also demonstrated for NO_2 sensing. This type of VG/CNT hybrid was created by the direct PECVD growth of VG nanosheets on a CNT surface. The increased surface area of the original CNT and the excellent electrical contact between the CNT and the VG in the hybrid favor gas sensing applications. This type of sensor showed enhanced sensing response to NO_2 compared with those of MWCNT and graphene/CNT mixture sensors (Fig. 6.5c). In addition to the enhanced sensing performance, the VG sensor is promising for gas sensing as the device is suitable for large-scale fabrication and has a better stability than sensors fabricated by other methods such as drop-casting of RGO sheets.

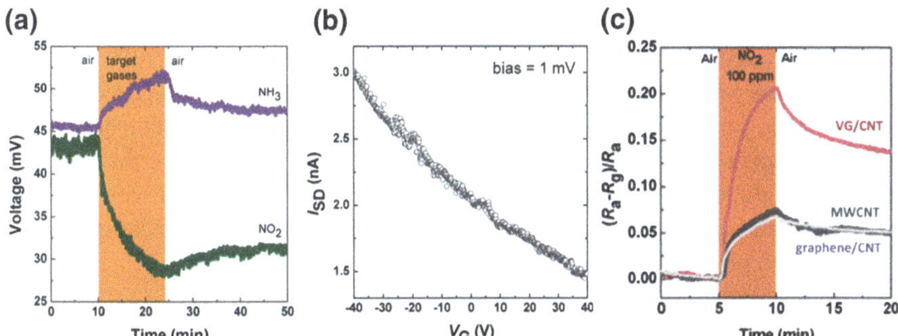

Fig. 6.5 **a** Room-temperature detection of NO_2 (100 ppm) and NH_3 (1 %) with the VG sensor. When NO_2 and NH_3 gases are injected in the testing chamber, the sensor conductivity shows significant changes. When NO_2 and NH_3 gases are shut off and an airflow is introduced, the sensor conductivity begins to recover. **b** The FET transport characteristic of a VG sensor. The sensor conductivity gradually decreases when the gate voltage changes from −40 to +40 V, showing that the VGs in the sensor behave like a p-type semiconductor. Reprinted with permission from [17]. Copyright 2011 American Chemical Society. **c** Room-temperature detection of NO_2 (100 ppm) with VG/CNT, MWCNT, and graphene/CNT sensors. Reprinted with permission from [18]. Copyright 2011 American Chemical Society

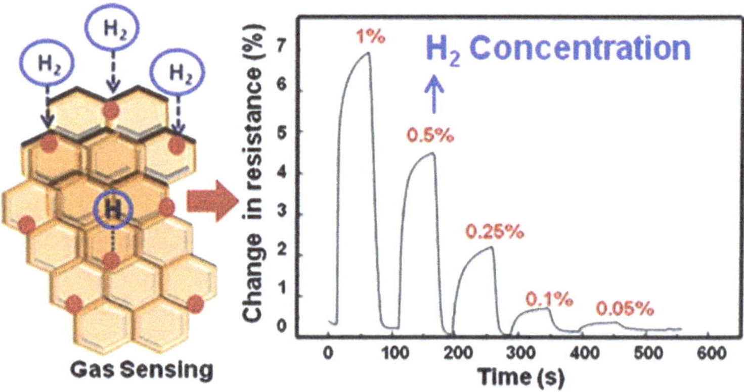

Fig. 6.6 Schematic of Pd–VG-based gas sensor and the hydrogen gas sensing results (ΔR/R vs. time). Reprinted with permission from [16]. Copyright 2013 Elsevier

VG shows good sensitivity to NO_2 and NH_3; however, the sp^2 carbon–carbon bonds in graphene are chemically stable, which leads to a relatively weak interaction between graphene and many other gas molecules, e.g., H_2. Functionalization of VG with noble metals, e.g., Pd and Pt, has been demonstrated to be an effective method to enhance the sensitivity of VG to these gas species. In one report, Pd NPs (10–20 nm) decorated VG sheets were investigated for hydrogen detection (Fig. 6.6) [16]. When hydrogen encounters Pd NP, it dissociates and forms palladium hydride, which donates electrons to the VG and decreases its conductivity. The sensor could detect hydrogen at low concentration (as low as 0.05 %) under room temperature with a fast response (12 s for 1 % hydrogen and 26 s for 0.1 % hydrogen) and could fully recover in 20 s. Theoretical studies have shown that the wrinkles/defects in a graphene sheet are more favorable for gas adsorption compared with other locations. Therefore, more gas adsorption sites are expected for VG due to the large number of wrinkles and edges, leading to higher sensitivities of the sensor compared with flat graphene-based devices. The demonstrated biosensor and gas sensor applications of VG rely on the unique properties from the vertical orientation of graphene sheets and the sensor performance could be further improved by decreasing the thickness of the VG sheet or engineering the VG structure with physical/chemical modifications.

Theoretical studies have shown that the graphene–gas molecule adsorption is strongly dependent on the graphene structure and the molecular adsorption configuration [19]. Gas molecules have much stronger adsorption on the doped or defective graphenes than on the pristine graphene. Therefore, high adsorption energy is expected for VGs with gas molecules due to a large number of defects, edges, and a curved morphology of the sheets, which lead to higher sensitivities of the sensor compared with horizontal graphene-based devices. The demonstrated biosensor and gas sensor applications of VGs rely on the unique properties arising from the vertical orientation of graphene sheets. In addition, the sensing performance could be further improved by decreasing the thickness of VG sheets or engineering the VG structure with physical/chemical modifications.

6.3 VG-Based Corona Discharge

Corona discharge is a localized breakdown phenomenon in gases, which is typically created by an asymmetric electrode pair (e.g., pin-to-plate and wire-to-plate). Corona discharges employing a microsized metallic wire as the discharge electrode are widely used for indoor electrostatic devices (e.g., photocopiers and printers). However, ozone is emitted as a hazardous by-product that poses serious health hazards to the human respiratory system.

A more health-benign corona discharge employing a VG-coated stainless steel (SS) wire as the discharge electrode was demonstrated (Fig. 6.7a, c) [20]. Due to the excellent electrical conductivity and a high aspect ratio, the electric field near the VG edges can easily reach a critical value for electrical breakdown. For negative discharge from a bare SS wire, the corona inception voltage was ~5.6 kV, at

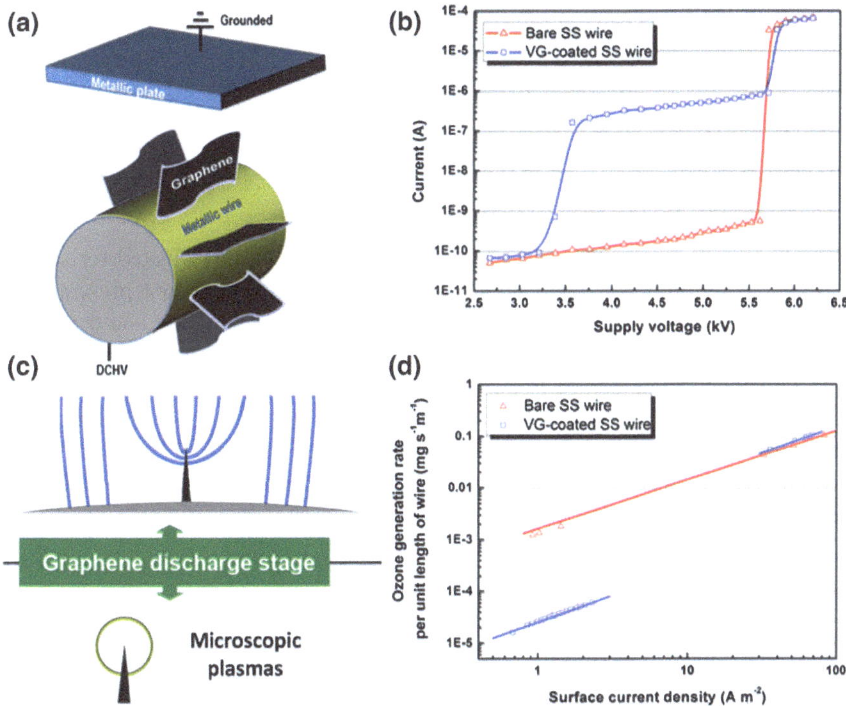

Fig. 6.7 **a** Schematic of a simplified configuration of VG-coated wire discharge. **b** I–V curves of negative corona discharges employing bare SS wire (*red line*) and VG-coated SS wire (*blue line*). Different from bare SS wire discharge showing a single exponential increase in the discharge current, the VG-coated SS wire discharge process includes two stages: 'graphene discharge stage' and 'wire discharge stage.' **c** Schematic of electric field lines and the plasma region for the VG-based corona discharge. **d** Coating of SS wires with VGs decreases the rates of ozone generation in the microscopic corona-type discharges by more than two orders of magnitude at a surface current density of <2 A m^{-2}. Reprinted with permission from [20]. Copyright 2011 The Royal Society of Chemistry

which the discharge current suddenly increased from 10^{-10} to 10^{-5} A (Fig. 6.7b, red line). In contrast, the VG-coated SS wire discharge exhibited quite different I–V characteristics (blue line). The I–V dependence of the VG-coated SS wire discharge could be divided into two discharge stages. In the first stage, the inception of corona discharge occurred as early as ~3.2 kV, with an exponential increase in corona current from 10^{-10} to 10^{-7} A. In the second stage, another inception of corona discharge occurred when the supply voltage was increased to ~5.6 kV, after which the I–V curve almost followed that of the bare SS discharge. As a consequence, a VG-based corona discharge can be initiated and operated at a much lower voltage as compared with common metal electrodes. More importantly, for a given surface current density (<2 A m^{-2}), the ozone generation rate per unit length of the wire in VG-based corona discharges is only 1 % of the amount of ozone produced using the microsized SS wire (Fig. 6.7d). Ozone generation in atmospheric-pressure corona discharges is affected by the size of the plasma region. Ionization is very effective in the plasma region, while electrons outside the plasma region are not energetic enough to generate ozone. Consequently, VG networks with ultrathin edges can generate microscopic corona discharges, further leading to much lower ozone emission rates.

6.4 Summary

VG, along with many other graphene materials, has been investigated for sensing and other environmental applications. In this chapter, we have first presented the representative VG-based sensors. Because of VG's unique structure and properties, VG structures have shown promise for bio- and gas sensing and can effectively detect various biomolecules (e.g., proteins, DNA, and bacteria) and gases (e.g., NO_2, NH_3, and H_2) at very low concentrations. As another important application of VG in environmental field, we have also introduced the use of VG as a 'green' corona discharge electrode with low ozone generation, which is attributed to the vertical orientation and the high conductivity of VG sheets.

References

1. Comini, E., Faglia, G., Sberveglieri, G., Pan, Z. W., & Wang, Z. L. (2002). Stable and highly sensitive gas sensors based on semiconducting oxide nanobelts. *Applied Physics Letters, 81*(10), 1869–1871.
2. Wan, Q., Li, Q. H., Chen, Y. J., Wang, T. H., He, X. L., Li, J. P., & Lin, C. L. (2004). Fabrication and ethanol sensing characteristics of ZnO nanowire gas sensors. *Applied Physics Letters, 84*(18), 3654–3656.
3. Eranna, G., Joshi, B. C., Runthala, D. P., & Gupta, R. P. (2004). Oxide materials for development of integrated gas sensors—A comprehensive review. *Critical Reviews in Solid State and Materials Sciences, 29*(3–4), 111–188.

4. Liu, Y., Dong, X., & Chen, P. (2012). Biological and chemical sensors based on graphene materials. *Chemical Society Reviews, 41*(6), 2283–2307.
5. Llobet, E. (2013). Gas sensors using carbon nanomaterials: A review. *Sensors and Actuators B-Chemical, 179*, 32–45.
6. Mao, S., Lu, G., & Chen, J. (2014). Nanocarbon-based gas sensors: Progress and challenges. *Journal of Materials Chemistry A, 2*(16), 5573–5579.
7. Bondavalli, P., Legagneux, P., & Pribat, D. (2009). Carbon nanotubes based transistors as gas sensors: State of the art and critical review. *Sensors and Actuators B-Chemical, 140*(1), 304–318.
8. He, Q., Wu, S., Yin, Z., & Zhang, H. (2012). Graphene-based electronic sensors. *Chemical Science, 3*(6), 1764–1772.
9. Mao, S., Lu, G., Yu, K., Bo, Z., & Chen, J. (2010). Specific protein detection using thermally reduced graphene oxide sheet decorated with gold nanoparticle-antibody conjugates. *Advanced Materials, 22*(32), 3521–3526.
10. Ratinac, K. R., Yang, W., Ringer, S. P., & Braet, F. (2010). Toward ubiquitous environmental gas sensors-capitalizing on the promise of graphene. *Environmental Science and Technology, 44*(4), 1167–1176.
11. Yavari, F., & Koratkar, N. (2012). Graphene-based chemical sensors. *Journal of Physical Chemistry Letters, 3*(13), 1746–1753.
12. Mao, S., Yu, K., Chang, J., Steeber, D. A., Ocola, L. E., & Chen, J. (2013). Direct growth of vertically-oriented graphene for field-effect transistor biosensor. *Scientific Reports, 3*, 1696.
13. Shang, N. G., Papakonstantinou, P., McMullan, M., Chu, M., Stamboulis, A., Potenza, A., et al. (2008). Catalyst-Free efficient growth, orientation and biosensing properties of multilayer graphene nanoflake films with sharp Edge planes. *Advanced Functional Materials, 18*(21), 3506–3514.
14. Akhavan, O., Ghaderi, E., & Rahighi, R. (2012). Toward single-DNA electrochemical biosensing by graphene nanowalls. *ACS Nano, 6*(4), 2904–2916.
15. Rider, A. E., Kumar, S., Furman, S. A., & Ostrikov, K. (2012). Self-organized Au nanoarrays on vertical graphenes: an advanced three-dimensional sensing platform. *Chemical Communications, 48*(21), 2659–2661.
16. Seo, D. H., Rider, A. E., Kumar, S., Randeniya, L. K., & Ostrikou, K. (2013). Vertical graphene gas-and bio-sensors via catalyst-free, reactive plasma reforming of natural honey. *Carbon, 60*, 221–228.
17. Yu, K., Wang, P., Lu, G., Chen, K.-H., Bo, Z., & Chen, J. (2011). Patterning vertically oriented graphene sheets for nanodevice applications. *Journal of Physical Chemistry Letters, 2*(6), 537–542.
18. Yu, K., Lu, G., Bo, Z., Mao, S., & Chen, J. (2011). Carbon nanotube with chemically bonded graphene leaves for electronic and optoelectronic applications. *Journal of Physical Chemistry Letters, 2*(13), 1556–1562.
19. Zhang, Y.-H., Chen, Y.-B., Zhou, K.-G., Liu, C.-H., Zeng, J., Zhang, H.-L., & Peng, Y. (2009). Improving gas sensing properties of graphene by introducing dopants and defects: A first-principles study. *Nanotechnology, 20*(18), 185504.
20. Bo, Z., Yu, K., Lu, G., Cui, S., Mao, S., & Chen, J. (2011). Vertically oriented graphene sheets grown on metallic wires for greener corona discharges: Lower power consumption and minimized ozone emission. *Energy and Environmental Science, 4*(7), 2525–2528.

Chapter 7
Vertically-Oriented Graphene for Supercapacitors

Abstract Energy storage devices not only are widely used in nearly all aspects of our daily lives, but also become an essential necessity for the development of renewable energy, which features intermittency. Many researchers around the world are working to improve the energy density, power density, safety, and charging speed of energy storage devices to meet the ever-increasing demands of storing electricity. In this context, extensive research has been and is being carried out on improving the performance of the two major electrochemical energy storage devices, namely batteries and supercapacitors. Traditional rechargeable batteries have shortcomings such as high cost, short life, long charging time, low power density, and potential environmental issues. On the other hand, electrochemical capacitors (i.e., the so-called supercapacitors or ultracapacitors) show great potential in energy storage because of excellent charge/discharge rates, long cycle life, and environmental friendliness. Vertically-oriented graphene (VG) has many characteristics that are ideal for building supercapacitors with improved performance and has been investigated as an electrode material for supercapacitors. In this chapter, supercapacitors using VG-based materials as electrodes are introduced and discussed.

Keywords Supercapacitor · Electric double-layer capacitor · *Pseudo*-capacitor

Energy storage devices can convert energy from forms that are difficult to store to more conveniently or economically storable forms. Such storage devices are not only widely used in all aspects of our daily lives, but also become an absolutely essential demand for the development of renewable energy. For example, the intermittent production of energy sources such as solar and wind power makes it hard to balance grid power and supply stable energy, thereby limiting their uses. Many researchers around the world focused on increasing the capacity, safety, and charging speed of energy storage devices to meet the demands of storing electricity.

Part of this chapter was adapted from our review article "Emerging Energy and Environmental Applications of Vertically-Oriented Graphenes," Chemical Society Reviews, 2015 (DOI: 10.1039/C4CS00352G)—Reproduced by permission of The Royal Society of Chemistry.

J. Chen et al., *Vertically-Oriented Graphene*, DOI 10.1007/978-3-319-15302-5_7

At present, the main electrochemical energy storage devices comprise batteries and electrochemical capacitors (i.e., the so-called supercapacitors or ultracapacitors). Traditional rechargeable batteries have many shortcomings, such as short life, long charging time, low power density, and potential environmental issues due to their chemical-based energy storage mechanism. On the other hand, supercapacitors present great potential in energy storage with higher charge–discharge rates, higher power density, longer cycle life, and environmental friendliness. Typically, compared with batteries, supercapacitors are able to get charged and discharged at relatively high rates, making supercapacitors very promising and attractive for a wide range of industrial applications. Since the energy storage mechanism of supercapacitors is the simple charge separation at the electrochemical interface between the electrode and the electrolyte, rapid charge/discharge, high power density, and long cycle life can be achieved without environmental and safety risks. Due to the use of advanced nanomaterials, such supercapacitor devices also have a much higher capacitance than regular capacitors. Nowadays, supercapacitors can be used either by themselves as a primary power source or in combination with batteries or fuel cells.

Similar to batteries and traditional capacitors, as shown in Fig. 7.1, supercapacitors have two electrodes that are charged reversely. The active materials in both electrodes are impregnated in an electrolyte together with the separator and are connected with current collectors. The separator can prevent electronic current from discharging the cell while allowing ionic current to flow between the electrodes. At the same time, the current collectors are used to conduct electrical current from active materials to the external circuit.

Supercapacitors can be basically divided into two categories: electric double-layer (EDL) capacitors and *pseudo*-capacitors. In EDL capacitors (EDLCs), electrons involved in double-layer charging are the delocalized conduction-band electrons and carbon-based materials are commonly used. On the other hand, metallic oxides and conducting polymers are the common active materials of *pseudo*-capacitors. In this chapter, recent progress of VGs and VG-based hybrids/composites for applications in both types of supercapacitors is introduced.

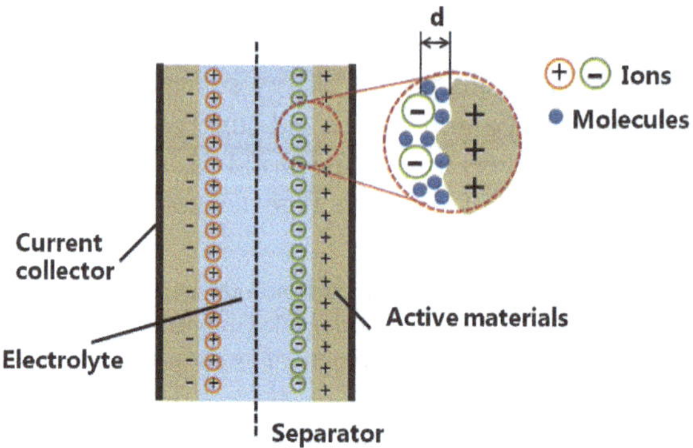

Fig. 7.1 Schematic of the basic structure of a supercapacitor

7.1 VG-Based Active Materials for EDLCs

Unlike rechargeable batteries that store energy through Faradaic redox reactions, EDLCs work in a direct, electrostatic way, where active materials are charged for the rapid separation and surface adsorption of ions with the opposite charge. EDLC consists of two array layers of opposite charges, separated by a small distance having atomic dimensions. The major advantages of EDLCs compared with other types of supercapacitors and rechargeable batteries are a higher power density, shorter charge/discharge cycles, and a longer life span. Besides the ultrathin charge separation distance (the thickness of EDL), the supercapacitance of EDLCs is mainly attributed to the use of active materials with a large surface area. Consequently, the characteristics of active materials play a crucial role in the performance of EDLCs.

As is well known, a high-performance EDLC active material should first have a high specific surface area. Because of their abundant raw sources, low cost, good formability, high electrochemical stability, and highly porous structure with a high specific surface area, carbon-based materials are widely used as active materials in EDLCs. Many types of carbon-based active materials have been proposed and tested, such as activated carbons (ACs), activated carbon fibers (ACFs), carbon aerogels (CAGs), CNTs, and graphene [1].

The most commonly used active materials for EDLCs are ACs, a class of porous structures that can be produced with a high specific surface area and a controlled pore structure. However, for EDLCs made from porous electrodes such as ACs, distributed charge storage will be formed with the so-called transmission line electrical response, leading to extra electrolyte resistance associated with the movement of ions within the pores. A similar problem also exists with using chemically modified graphene (CMG) sheets as the active materials, where the pores mainly originate from the graphene interlayer spacings. Due to the van der Waals interactions and the use of binders, easy restacking of CMG nanosheets is usually observed, generating obvious ionic resistance within agglomerated graphene layers. With the above porosity effect, typical AC- and CMG-based EDLCs commonly present a resistor–capacitor (RC) time constant of ~1 s [2]. Unfortunately, this value cannot meet the requirements of 120-Hz ac line-filtering applications (smooth transition from ac to dc, 8.3 ms in period) for line-powered electronics.

Besides the high surface area, an ideal EDLC active material should fulfill the following requirements [3]:

(i) A high electrolyte-accessible surface area for the effective adsorption of ions to obtain a high capacitance and a high energy density;
(ii) A suitable structure for the easy diffusion of ions to realize high rate capability; and
(iii) Minimum resistances within the material and at the contact interface of material/current collector for the fast electron transport to obtain a high power density.

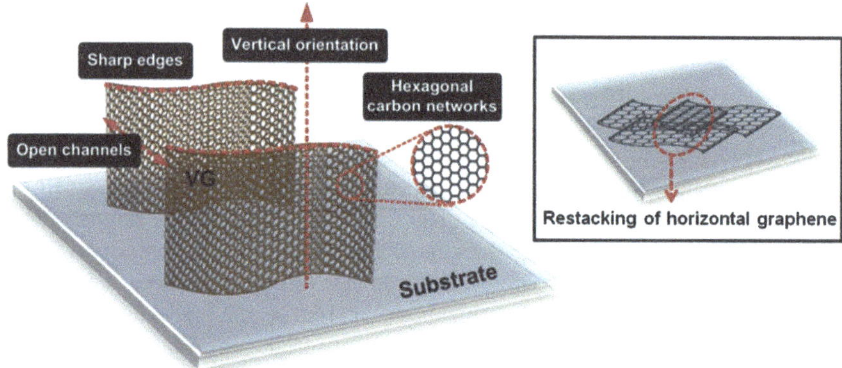

Fig. 7.2 A schematic representing VGs' structural and morphological features. Inset illustrates the restacking of horizontal graphene nanosheets [3]. Copyright 2015 The Royal Society of Chemistry

The unique features of VGs enable them to ideally satisfy the above criteria. First, VG networks present a non-agglomerated structure with exposed edge planes, which facilitates the surface use for charge storage, as shown in Fig. 7.2. The conventional graphene nanosheets are almost parallel to the current collector, and thus, the structure can be seen as the horizontal graphene stacks. Due to van der Waals interactions and the use of binders, the commonly observed restacking of the horizontal graphene nanosheets leads to a considerable reduction of the available surface areas for charge adsorption and electrochemical reactions. In contrast, VG's non-agglomerated morphology leads to a higher electrochemically accessible surface area, and thus to a higher capacitance [3]. Meanwhile, the dense edge planes of VGs can also enhance the charge storage capability, since the edge planes have a much larger area-specific capacitance (50–70 $\mu F/cm^2$) than the basal plane surface, which provides area-specific capacitance of only about 3 $\mu F/cm^2$ [2].

Second, when VGs are used as active materials, the large ionic resistance associated with the distributed charge storage in porous materials can be minimized, making it possible to use EDLCs in a high-frequency mode [3]. Electrolyte access into the pores plays a major role in EDLC's rate capability and frequency response. Previous research showed that the pore size leading to the maximum double-layer capacitance is very close to the ion size, and both larger and smaller pores can lead to a significant drop in capacitance [4]. Considerable ionic resistance is formed when the pores are small and tortuous, thus leading to a poor capacitive behavior, especially at relatively high frequencies [2]. This problem not only exists for activated carbons (ACs, the most common porous active materials for commercial EDLCs), but also for horizontal graphene stacks where pores mainly originate from the 2-D interlayer spacings. As for VGs (see Fig. 7.3), the vertical orientation and open intersheet channels are exposed and directly

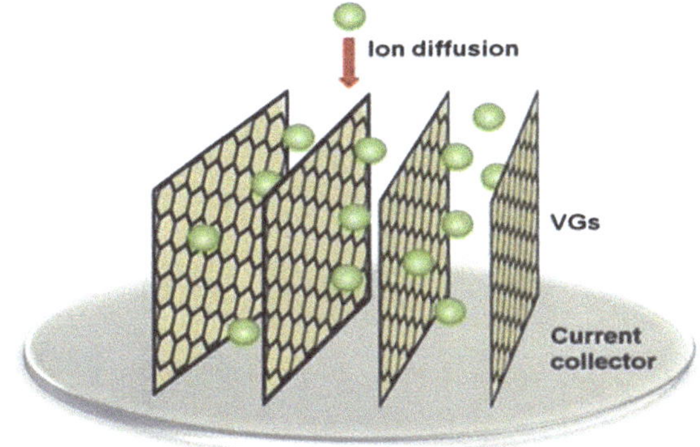

Fig. 7.3 Schematic of ion diffusion within VGs [3]. Copyright 2015 The Royal Society of Chemistry

accessible without porosity; thus, they can facilitate the ion migration between the layers and minimize the undesirable porosity effects, especially at high frequencies.

Third, the EDLC series resistance can be reduced and the power and rate capabilities can be significantly enhanced by using VGs. Since the in-plane electrical conductivity is much higher than that of out of plane, the vertical orientation facilitates the charge transport within active materials. In addition, the direct growth of VGs without a commonly used binder can reduce the contact resistance between active materials and the current collector.

A significant feature of VG-based EDLCs is the ultrafast dynamic response. In particular, the 120-Hz ac line filtering has become possible [2]. The smooth transition from 120 Hz ac to dc is required for line-powered electronics and relies on traditional electrolytic capacitors. Since aluminum electrolytic capacitors are usually the largest components in electronic devices, replacing them by more compact capacitors has an impact on future electronic devices. A variety of materials, including active carbon [5], CNTs, [6] carbon black [7], and horizontal graphene stacks [8], have been tested as EDLC electrodes for ac line filtering. Although EDLCs can provide a much higher specific capacitance, most devices made from porous materials fail to show the capacitive behavior at relatively high frequencies, which is due to the porosity effect discussed above [2]. In 2010, J.R. Miller and coworkers reported the ac line-filtering application of VG-based EDLCs [2]. VGs were synthesized directly on nickel substrates via a radio-frequency plasma-enhanced chemical vapor deposition (rf-PECVD) reactor. The substrates were first plasma etched in a mixture of hydrogen and argon. Subsequently, methane was used as the carbon source for VG growth. The VG layer is 600 nm in height and <1 nm in the thickness of each nanosheet. Figure 7.4a shows the SEM image of the typical VGs grown on Ni substrates. As expected, the VG networks presented exposed edge planes and open channels between adjacent nanosheets. The surface area of VGs was reported as ~1000 m^2/g. A VG-based EDLC was fabricated with KOH as the electrolyte for electrochemical tests. As shown in Fig. 7.4b, VG-based EDLC presented an almost vertical Nyquist plot, which means that the capacitive impedance is totally dependent on 1/C. Meanwhile, no features of either ionic resistance or series resistance

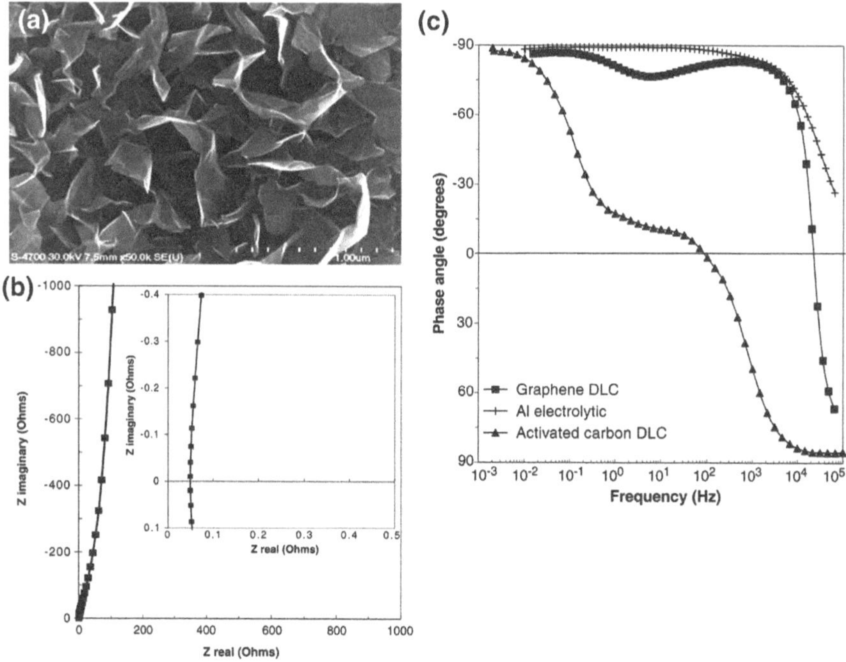

Fig. 7.4 VG-based EDLC for ac line filtering. **a** SEM micrograph of VGs on the Ni electrode. **b** A complex plane plot of the impedance data obtained from VG-based EDLC. **c** Impedance phase angle data of a VG-based EDLC (labeled as 'Graphene DLC' in the figure), an AC-based EDLC (labeled as 'Activated carbon DLC' in the figure), and an Al electrolytic capacitor. Reprinted with permission from [2]. Copyright 2010 The American Association for the Advancement of Science

were observed. As shown in Fig. 7.4c, AC-based EDLC features an impedance phase angle of ~0° at 120 Hz (behaves like a resistor). In contrast, the VG-based EDLC exhibited an impedance phase angle of −82° at the same frequencies, which resembles commonly used Al electrolytic capacitors and is very close to the ideal capacitive behavior (−90°). This indicates that VG-based EDLC is suitable for the 120-Hz ac line-filtering application. Meanwhile, compared with the Al electrolytic capacitor owning an impedance phase angle of −83° at 120 Hz, the VG-based EDLC presents a much higher volumetric energy density and thus allows for size reductions of the filtering system. The 0.6-μm-thick VG layer stored ~1.5 and ~5.5 FV/cm^3 with the aqueous and organic electrolytes, respectively, significantly higher than that of an Al electrolytic capacitor (~0.14 FV/cm^3) [2].

Further, kHz ultrafast EDLCs, which can work at a frequency 3–4 orders higher than traditional EDLCs, have been demonstrated. As reported by Ren et al. [9], VGs were grown on nickel foam current collectors using a microwave PECVD reactor. As shown in Fig. 7.5, VGs were deposited everywhere on the Ni skeleton. The close view of the morphology of VGs can be found in the SEM image shown in Fig. 7.5a, b. The use of a foam-type current collector instead of a foil-type counterpart can lead to a higher mass loading of active materials, as VGs can fully cover the 3-D metallic scaffold, as schematically shown in Fig. 7.5c.

By combining the large surface area with the straightforward porous structure, fast ion migration and huge capacitance are ensured. Electrochemical measurements of such a VG-based EDLC were conducted using 6 M KOH as the electrolyte. Cyclic voltammetry (CV) curves at several scan rates from 1 to 500 V/s of a VG-based EDLC are shown in Fig. 7.6a–e. The shape of the CV curves was found to be quasi-rectangle for all the tested scan rates, even at an ultrahigh value of 500 V/s, suggesting the typical EDLC behavior with excellent charge propagation at electrode and electrolyte interfaces. It is noted that reports

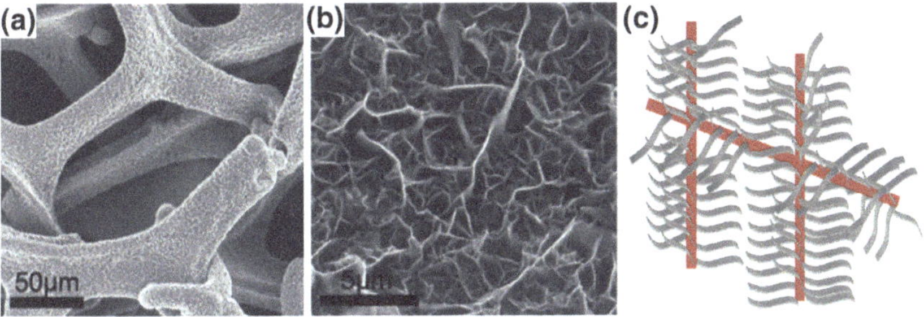

Fig. 7.5 VGs grown on the nickel foam for kilohertz ultrafast EDLCs. **a**–**b** SEM micrographs of POG/NF electrode under different magnifications. **c** 3-D network of graphene grass. Reprinted with permission from [9]. Copyright 2014 Elsevier

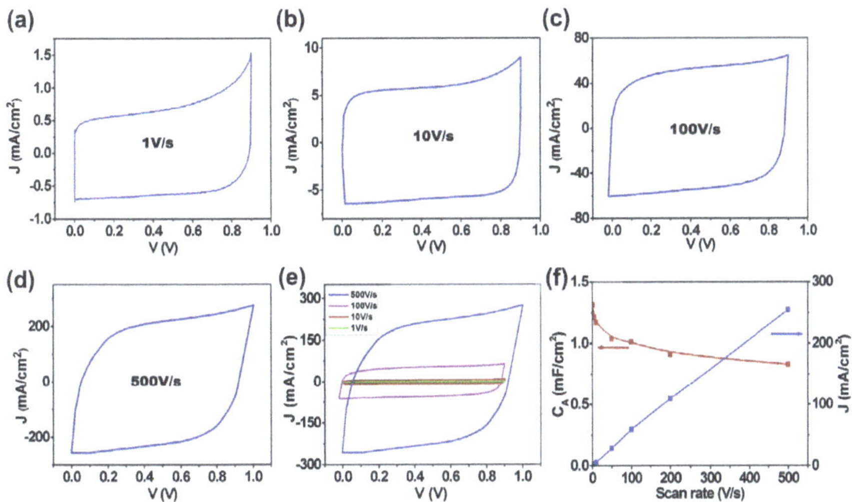

Fig. 7.6 **a**–**e** Cyclic voltammetry curves of the cells at scan rates from 1 V/s to 500 V/s in 6 M KOH. **f** The discharge current density and the calculated electrode-specific capacitance at different scan rates. Reprinted with permission from [9]. Copyright 2014 Elsevier

on most EDLC studies are conducted using the highest scan rate of 1 V/s. Figure 7.6f shows the variation of the discharge current density and the calculated electrode-specific capacitance at different CV scan rates. The almost linear relationship between the discharge current and the scan rate, as well as the good capacitance retention confirms the high rate performance of such a VG-based EDLC. Meanwhile, the proposed VG-based EDLCs exhibited a capacitance of ~0.32 mF cm^{-2} at 1 kHz, higher than any previously reported values of EDLCs at the same frequency [9].

The capacitive behaviors of VGs can be tailored by tuning their morphology and structure in the PECVD growth. For example, the density of edge planes and the graphitization degree strongly affect their electrochemical performance [10]. Specifically, thinner edge planes lead to a higher specific capacitance [10], arising from the much larger area-specific capacitance of edge planes than basal surface planes [2]. Meanwhile, a higher sp^2 content also improves the charge storage capability, since the sp^3-bonded carbon only increases the charge transfer resistance and contributes little to the charge storage [10]. K. Ostrikov's group demonstrated the direct fabrication of VGs on porous nickel foam by using cheap and spreadable natural fatty precursors [10]. Controllable VG structures were obtained by adjusting the growth parameters, such as the H_2 concentration in the feeding gases. They found that the graphitic ordering increased along with the increase in H_2 concentration, while the thickness of edge planes decreased, as shown in the SEM images (Fig. 7.7a–c) and the Raman spectra (Fig. 7.7d). Figure 7.7e, f shows the CV and galvanostatic charge/discharge curves of VGS grown with 80 % H_2. For this typical sample, the specific capacitance of VG-based EDLCs could reach a high value of 230 F/g at a CV scan rate of 10 mV/s (shown in Fig. 7.7e–f) [10]. Finally, a high capacitance retention of >99 % after 1000 cycles at a scan rate of 100 mV/s was observed, indicating the excellent stability of the tested electrode.

The specific capacitance can also be improved by using the VG-based hybrid structures, e.g., CNT-on-VG [11]. Figure 7.8a illustrates the procedure to prepare the CNT-on-VG electrode. Plasma was used to break down the carbon-containing molecules in butter (precursor) and reconstruct them into ordered and vertical graphitic structures [11]. Subsequently, CNTs were grown on the surface of VG nanosheets with Co/Mo as the catalysts. The SEM images of the as-grown pristine VGs and CNT-on-VG hybrid structure are presented in Fig. 7.8b–d.

Figure 7.9 shows the CV and galvanostatic charge/discharge curves of CNT-on-VG, pristine VGs, and pure CNTs. It was found that such a combination of 1-D and 2-D nanostructures in a CNT-on-VG hybrid structure can further increase the surface area and enhance the electron transport within active materials, leading to improved capacitive performance. Typically, the CNT-on-VG hybrid structure presented a high specific capacitance (278 F/g at 10 mV/s) and a good cycling stability (capacitance retention of >99 % after 8000 charge/discharge cycles) [11].

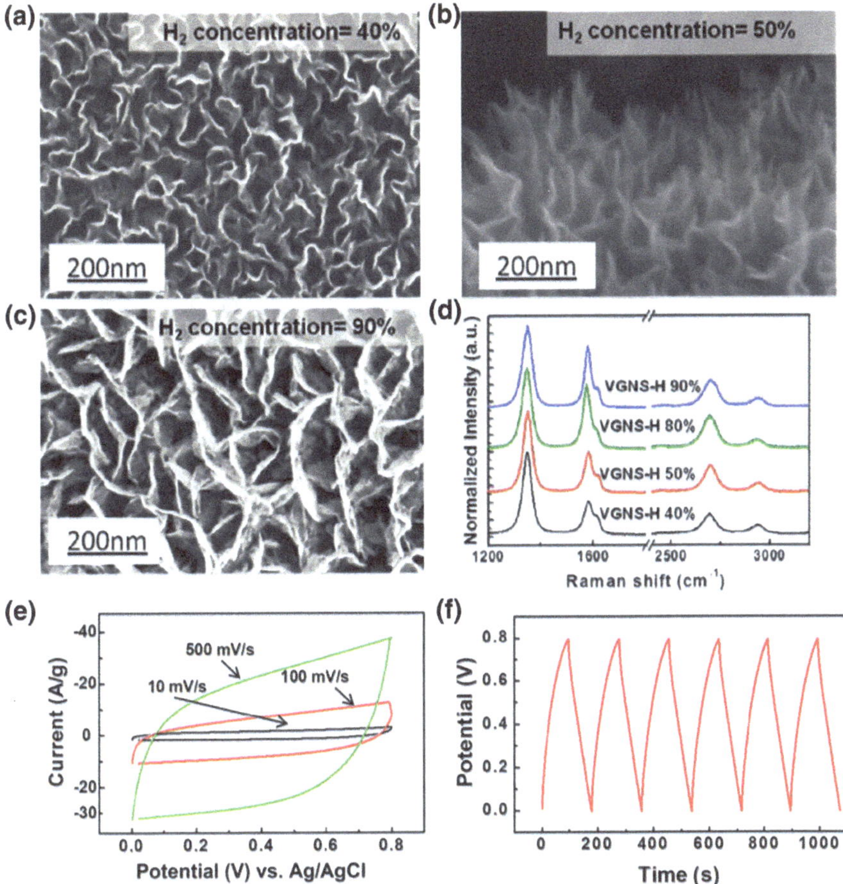

Fig. 7.7 Graphitic structure- and edge plane-controlled VGNS and optimized performance. **a**–**c** The SEM images of VGNS grown with 40, 50, and 90 % H_2. **d** The Raman spectra of VGNS grown with varying H_2 concentrations. **e** CV curves of VGNS grown with 80 % H_2 at varying scan rates. **f** Galvanostatic charge/discharge curves of VGNS grown with 80 % H_2. Reprinted with permission from [10]. Copyright 2013 John Wiley and Sons

7.2 VG-Based Active Materials for Pseudo-Capacitors

In contrast to EDLCs based on the physical adsorption of ions, *pseudo*-capacitors (also known as redox supercapacitors) rely on *pseudo*-capacitance derived from reversible Faradaic-type charge transfer in electrodes. Due to the presence of Faradaic redox reactions between the electrolyte and electroactive species on the electrode surface, *pseudo*-capacitors usually have a higher capacitance. These electroactive species (referred to herein as *pseudo*-species) enables repeated Faradaic-type reactions. Traditional *pseudo*-species such as ruthenium oxide, manganese oxide, vanadium nitride, electrically conducting polymers, and

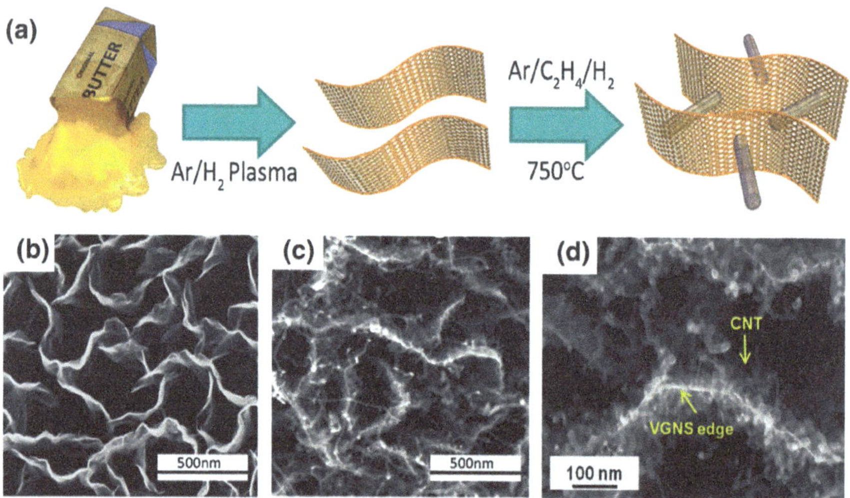

Fig. 7.8 CNT-on-VG **a** Schematic for the direct growth of CNTs onto VGNS. **b** SEM micrograph of pristine VGNS prior to CNT growth. **c** SEM micrograph of the final hybrid VGNS/CNT nanoarchitecture in which the graphene nanosheets were decorated with a high density of CNTs. **d** High-resolution SEM images of the VGNS/CNTs hybrid structure. Reprinted with permission from [11]. Copyright 2014 John Wiley and Sons

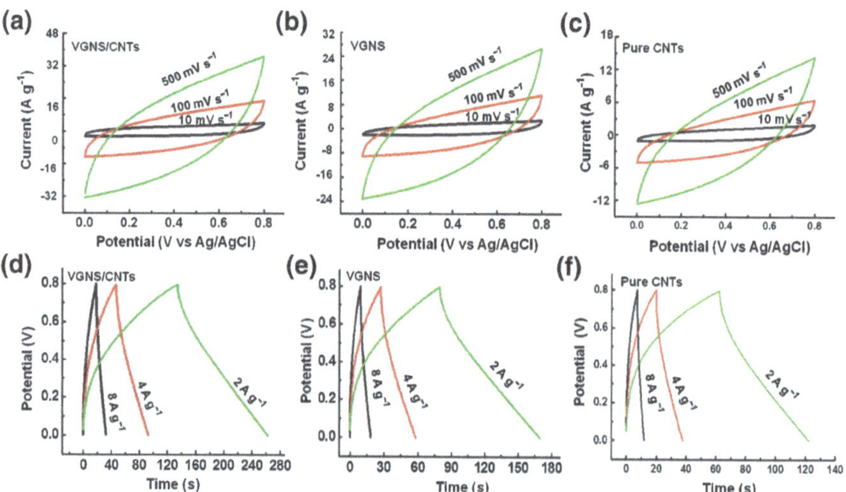

Fig. 7.9 CV curves of **a** CNT-on-VG, **b** pristine VGs, and **c** pure CNTs at scan rates of 10, 100, and 500 mV/s. Galvanostatic charge/discharge plots of **d** CNT-on-VG, **e** pristine VGs, and **f** pure CNTs at current densities of 2, 4, and 8 A/g. Reprinted with permission from [11]. Copyright 2014 John Wiley and Sons

oxygen- or nitrogen-containing surface functional groups are commonly used in *pseudo*-capacitors [1]. Due to the mechanism of chemical reactions instead of rapid separation and surface adsorption, *pseudo*-capacitors usually have a lower power density and poorer cycling stability compared with EDLC devices.

By combining merits of VGs and *pseudo*-species, high-performance *pseudo*-capacitors can be realized. VGs with both the high specific surface area and the high electrical conductivity can synergistically enhance the electrochemical performance of *pseudo*-species. Indeed, the roles of VGs in *pseudo*-capacitors mainly include the following [3]:

(i) Increasing the specific loading of the *pseudo*-species for a higher energy density,
(ii) Enhancing the charge transport between the *pseudo*-species and the substrate for higher power and rate capabilities, and
(iii) Improving the adhesion of the *pseudo*-species for the enhanced cycling stability.

As mentioned above, separation and adsorption also exist in *pseudo*-capacitors and the presence of VGs themselves can increase the specific surface area, thus contributing to the enhancement of capacitance.

As an example, electrochemical properties of *pseudo*-capacitors based on hybrid MnO_2/VG nanoarchitectures can compete with those of EDLCs [10]. In this work, VGs grown on a nickel foil were used as conductive templates for the deposition of MnO_2 nanoflowers. The exposed surface and high conductivity of VGs can enhance the electrochemical properties of MnO_2. Compared with pure VGs, the capacitance of the MnO_2/VG electrode was significantly enhanced due to the introduction of *pseudo*-species. At a CV scan rate of 10 mV/s, the MnO_2/VG electrode presented a high specific capacitance of 1060 F/g (calculated based on the mass of MnO_2, the CV curves are shown in Fig. 7.10a). Moreover, the

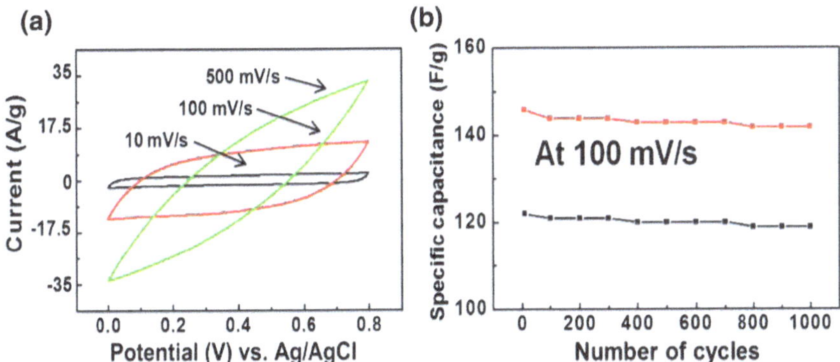

Fig. 7.10 **a** CV curves of the hybrid MnO_2/VGs grown with 80 % H_2 at varying scan rates. **b** The specific capacitance and cycle stability of MnO_2/VGs hybrids grown with 40 % (*black*) and 80 % (*red*) H_2. Reprinted with permission from [10]. Copyright 2013 John Wiley and Sons

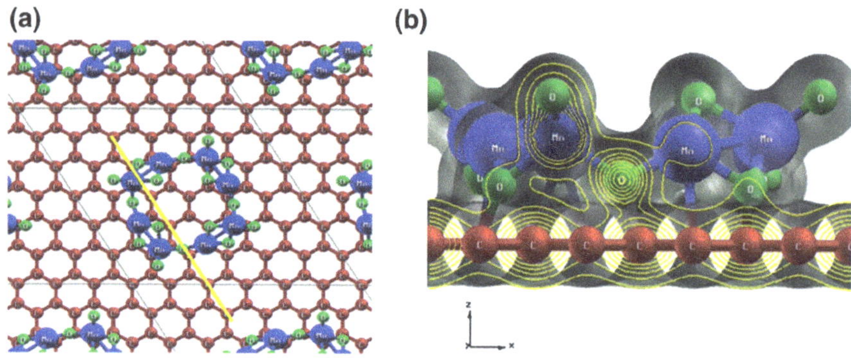

Fig. 7.11 **a** Schematic diagram of MnO_2 clusters and graphene. **b** Isoelectronic charge contour plot shown at a particular plane. Reprinted with permission from [12]. Copyright 2013 Elsevier

MnO_2/VG electrode exhibited a remarkable capacitance retention (>97 %) after 1000 cycles (Fig. 7.11b), which can be attributed to the strong adhesion between the MnO_2 and VGs [10].

Xiong et al. conducted density functional theory (DFT) calculations to elucidate the fundamental properties at the MnO_2/graphene interface. Based on the DFT calculations, the vertical orientation and sharp edge planes of VGs facilitate the ion diffusion with low energy barriers, while the covalent bonding between MnO_2 and graphene leads to the effective charge transfer [12]. Figure 7.11a shows the schematic diagram of MnO_2 clusters and graphene. Only one layer of graphene and a simple MnO_2 cluster were considered. The isoelectronic charge contour plot shown in Fig. 7.11b is a two-dimensional cut of the charge density in a vertical plane that contains the yellow line drawn parallel to the zigzag direction. These plots might indicate that the metallic nature of the MnO_2/graphene composite provides a facile conduction path for electron transport in the charge/discharge process.

Similar applications of VGs decorated with MnO_2 of diverse morphologies (such as VGs grown on flexible commercial buckpaper) [12] and other transition metal oxides (such as NiO) [13] have also been reported. These studies demonstrate the critical roles of VGs in the reduction of internal resistance, the enhancement of specific capacitance, and the improvement of cyclic stability.

VGs can also work with electrically conducting polymers for *pseudo*-capacitors. Xiong et al. demonstrated that hierarchical electrodes composed of carbon cloth (CC), VGs, and electrically conducting polyaniline (PANI) can further improve the supercapacitor performance [14]. The CC/VGs/PANI electrode is schematically shown in Fig. 7.12a. VGs were grown on the surface of CC by a microwave PECVD process, where CC served as a flexible and open scaffold. Then, the PANI nanoscale thin layer was coated on CC/VGs by an electropolymerization method. It was found that the specific surface area of CC increased by a factor of ~3 with the decoration of VGs. Figure 7.12b shows the CV curves of

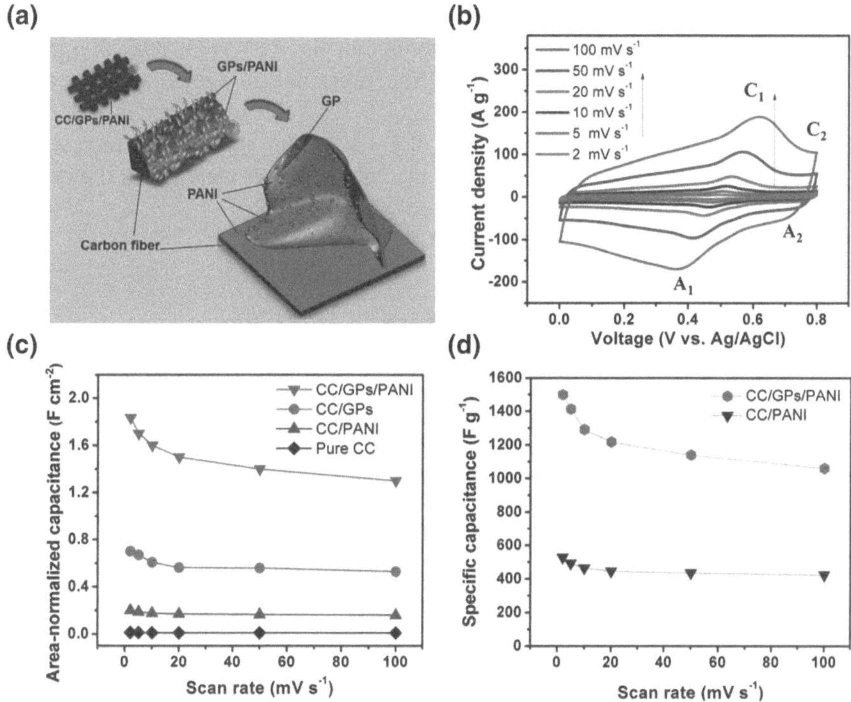

Fig. 7.12 **a** A hierarchical electrode composed of CC, VG (labeled as 'GP' in the figure), and PANI, for *pseudo*-capacitors. **b** CV curves for a hybrid CC/VGs/PANI composite electrode at scan rates of 2, 5, 10, 20, 50, and 100 mV/s in 1 M H_2SO_4 aqueous electrolyte. **c** Comparison of area-normalized specific capacitance of pure CC, CC/VGs, CC/PANI, and CC/VGs/PANI electrodes at different scan rates. **d** Comparison of mass-specific capacitance of PANI on both pure CC and CC/GP substrates. Reprinted with permission from [14]. Copyright 2013 John Wiley and Sons

CC/VGs/PANI (5 min of PANI electropolymerization) at scan rates of 2, 5, 10, 20, 50, and 100 mV/s with potential windows ranging from 0 to 0.8 V versus Ag/AgCl in 1 M H_2SO_4 aqueous electrolyte. Redox peaks were observed, indicating the Faradaic transformation of emeraldine–pernigraniline and the formation/reduction of bipolaronic pernigraniline [14]. Figure 7.12c presents the area-normalized specific capacitances of pure CC, CC/VGs, CC/PANI, and CC/VGs/PANI at different scan rates. The CC/VGs/PANI electrode presented a mass-specific capacitance of 2000 F/g and an area-normalized specific capacitance of 2.6 F/cm. Figure 7.12d displays a comparison of mass-specific capacitance of CC/PANI and CC/VGs/PANI. At a scan rate of 2 mV/s, the mass-specific capacitance of CC/VGs/PANI was also ~3 times higher compared with that of the CC/PANI electrode.

The presence of VGs also enhanced the electron transport, leading to energy and power densities higher than the previously reported values for PANI-based counterparts. A high-performance all-solid-state flexible supercapacitor based on the CC/VGs/PANI electrode and the polyvinyl alcohol (PVA)-H_2SO_4 polymer gel electrolyte was also demonstrated [14]. The supercapacitor presented an energy density ~10 times higher than a commercial 3.5 V/25 mF supercapacitor and comparable with the upper range of that of a 4 V lithium thin-film battery, yet the power density of which was ~2 orders of magnitude higher than that of the lithium thin-film battery.

7.3 VGs for Bridging Active Materials and Current Collectors

VGs are also used as bridges connecting active materials and current collectors in supercapacitors for the fast transport of electrons. Due to the manufacturing process and the nature of the materials, neither the current collector surface nor the active material surface is ideally flat. At the microscale, they meet at a finite number of contact points, thus leading to a considerable contact resistance at the interface. According to Holm's theory, the constriction/spreading resistance is highly dependent on the electrical resistivity of the contact spots. Graphene with superior electrical conductivity is thus attractive to use as an ideal bridge to offer high-quality electrical links at the interface between the current collector and the active material. As schematically shown in Fig. 7.13, VG bridges with an excellent in-plane electrical conductivity can remarkably reduce the contact resistance between the current collector and the active materials, leading to supercapacitors with high rate performance and a high power density [15].

Even with a conventional graphene film used as an active material in this proof-of-concept device, the VG-bridged supercapacitors outperformed nearly all of the previously reported counterparts. Four types of working electrodes were fabricated for comparison, i.e., without a current collector (no-CC), with a nickel foil current collector (foil-CC), with a nickel foam current collector (foam-CC), and

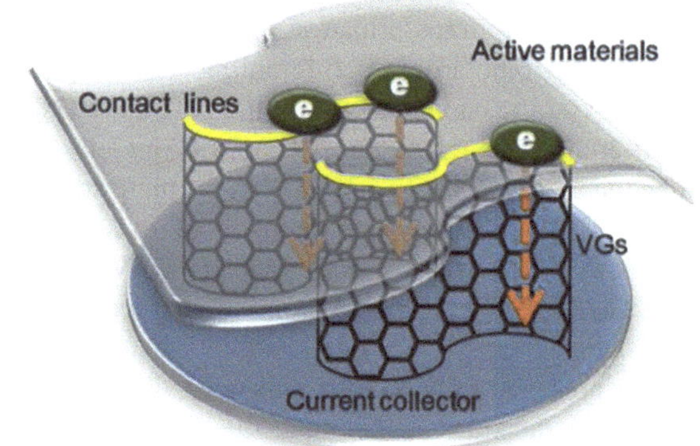

Fig. 7.13 VG bridges connecting active materials and a current collector increase contact area and enhance charge transport. Reprinted with permission from [15]. Copyright 2013 John Wiley and Sons

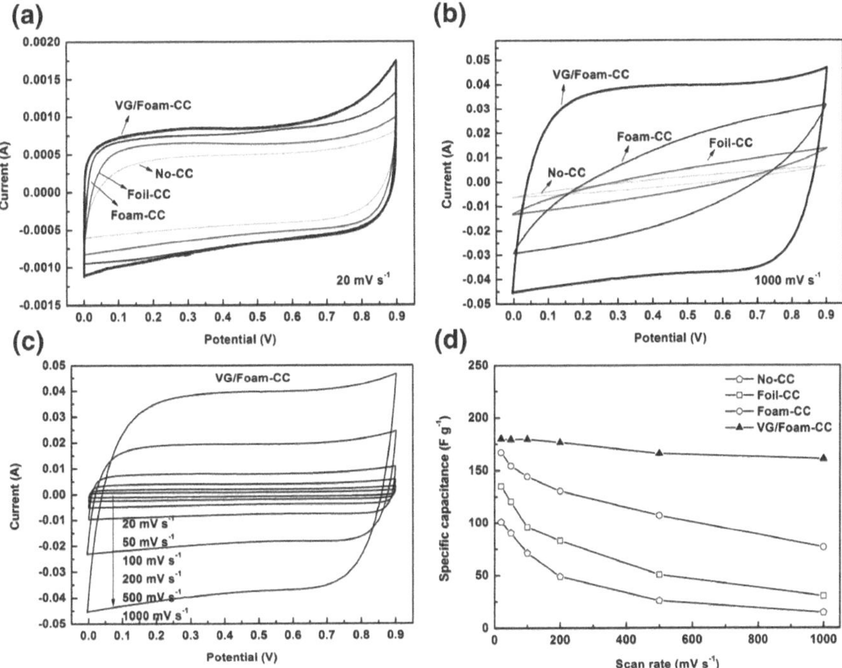

Fig. 7.14 CVs of different working electrodes at a scan rate of **a** 20 and **b** 1000 mV/s. **c** CVs of VG/foam-CC supercapacitor at different scan rates between 20 and 1000 mV/s. **d** Specific capacitance dependence on potential sweep rates for supercapacitors using different working electrodes (Electrolyte: 6.0 M KOH aqueous solution). Reprinted with permission from [15]. Copyright 2013 John Wiley and Sons

with a VG-coated nickel foam current collector (VG/foam-CC). According to the electrochemical impedance spectroscopy (EIS) tests, the characteristic relaxation time constant (CRTC) of the VG/foam-CC electrode was 32.8 ms, significantly lower than those of the no-CC (1538.5 ms), foil-CC (943.4 ms), and foam-CC (362.3 ms) electrodes. Figure 7.14a, b shows the CV curves of four types of working electrodes at scan rates of 20 and 1000 mV/s, respectively. With an increasing scan rate, the VG/foam-CC also manifested an enhanced area of CV curve but maintained the quasi-rectangular shape very well, as shown in Fig. 7.14c. Notably, the VG/foam-CC retained a specific capacitance of 160.7 F/g at a high scan rate of 1000 mV/s, which was decreased by only 10.7 % in specific capacitance compared with the value at a scan rate of 20 mV/s. This result was further confirmed by the galvanostatic charge/discharge tests. With the increase in galvanostatic charge/discharge, the current density increased from 1 to 100 A/g and the capacitance retention of VG/foam-CC electrode was found as high as ~90 % [15].

The exposed edges of VGs can provide dense contact points with active materials, thereby reducing the contact resistance. Meanwhile, due to the high in-plane electrical conductivity of graphene nanosheets, the vertical orientation of VGs enables the high-quality electrical contact at the interface between the current collector and the active material. Otherwise, inserting horizontal graphene films instead of VGs may cause a negative effect on the rate capability due to the increase of the internal resistance between planar graphene sheets [15]. This type of application of VGs could also be suitable for other electrochemical energy storage and conversion devices to advance their performance.

7.4 Summary

Supercapacitors, also known as electrochemical capacitors or ultracapacitors, have great potential in energy storage because of their excellent charge/discharge rates, long cycle life, and environmental friendliness. In this chapter, we have discussed the application of VG-based materials in two types of supercapacitors, namely EDLCs and *pseudo*-capacitors. VG has many characteristics ideal for building supercapacitors with improved performance. VG is an ideal EDLC active material in view of its high surface area accessible by electrolyte, a vertical structure suitable for the easy diffusion of ions, its high conductivity, and the minimum resistance at the contact interface of material/current collector. On the other hand, high-performance *pseudo*-capacitors can be fabricated by combining merits of VGs and pseudo-species. We have also demonstrated the use of VG sheets as bridges connecting active materials and current collectors in supercapacitors for the fast transport of electrons.

References

1. Zhang, L. L., & Zhao, X. S. (2009). Carbon-based materials as supercapacitor electrodes. *Chemical Society Reviews, 38*(9), 2520–2531.
2. Miller, J. R., Outlaw, R. A., & Holloway, B. C. (2010). Graphene double-layer capacitor with ac line-filtering performance. *Science, 329*(5999), 1637–1639.
3. Bo, Z., Mao, S., Han, Z. J., Cen, K., Chen, J., & Ostrikov, K. (2015). Emerging energy and environmental applications of vertically-oriented graphenes. *Chemical Society Reviews*, doi:10.1039/C4CS00352G.
4. Largeot, C., Portet, C., Chmiola, J., Taberna, P.-L., Gogotsi, Y., & Simon, P. (2008). Relation between the ion size and pore size for an electric double-layer capacitor. *Journal of the American Chemical Society, 130*(9), 2730–2731.
5. Feng, J., Sun, X., Wu, C., Peng, L., Lin, C., Hu, S., et al. (2011). Metallic few-layered VS2 ultrathin nanosheets: high two-dimensional conductivity for in-plane supercapacitors. *Journal of the American Chemical Society, 133*(44), 17832–17838.
6. Futaba, D. N., Hata, K., Yamada, T., Hiraoka, T., Hayamizu, Y., Kakudate, Y., et al. (2006). Shape-engineerable and highly densely packed single-walled carbon nanotubes and their application as super-capacitor electrodes. *Nature Materials, 5*(12), 987–994.
7. Kossyrev, P. (2012). Carbon black supercapacitors employing thin electrodes. *Journal of Power Sources, 201*, 347–352.

8. Sheng, K., Sun, Y., Li, C., Yuan, W., & Shi, G. (2012). Ultrahigh-rate supercapacitors based on eletrochemically reduced graphene oxide for ac line-filtering. *Scientific Reports, 2*, 247.
9. Ren, G., Pan, X., Bayne, S., & Fan, Z. (2014). Kilohertz ultrafast electrochemical supercapacitors based on perpendicularly-oriented graphene grown inside of nickel foam. *Carbon, 71*, 94–101.
10. Seo, D. H., Han, Z. J., Kumar, S., & Ostrikov, K. (2013). Structure-controlled, vertical graphene-based, binder-free electrodes from plasma-reformed butter enhance supercapacitor performance. *Advanced Energy Materials, 3*(10), 1316–1323.
11. Seo, D. H., Yick, S., Han, Z. J., Fang, J. H., & Ostrikov, K. (2014). Synergistic fusion of vertical graphene nanosheets and carbon nanotubes for high-performance supercapacitor electrodes. *ChemSusChem, 7*, 2317–2324.
12. Xiong, G., Hembram, K. P. S. S., Reifenberger, R. G., & Fisher, T. S. (2013). MnO_2-coated graphitic petals for supercapacitor electrodes. *Journal of Power Sources, 227*, 254–259.
13. Chang, H.-C., Chang, H.-Y., Su, W.-J., Lee, K.-Y., & Shih, W.-C. (2012). Preparation and electrochemical characterization of NiO nanostructure-carbon nanowall composites grown on carbon cloth. *Applied Surface Science, 258*(22), 8599–8602.
14. Xiong, G., Meng, C., Reifenberger, R. G., Irazoqui, P. P., & Fisher, T. S. (2014). Graphitic petal electrodes for all-solid-state flexible supercapacitors. *Advanced Energy Materials, 4*(3), 1300515.
15. Bo, Z., Zhu, W., Ma, W., Wen, Z., Shuai, X., Chen, J., et al. (2013). Vertically oriented graphene bridging active-layer/current-collector interface for ultrahigh rate supercapacitors. *Advanced Materials, 25*(40), 5799–5806.

Chapter 8
Vertically-Oriented Graphene for Other Energy Storage and Conversion Applications

Abstract Vertically-oriented graphenes (VGs) have outstanding electrical conductivity, robust binding with the substrate, and high chemical tolerance in electrolytes. These features of VGs offer great promise in electrochemical energy storage and conversion devices. Following our discussion in the previous chapter on the application of VGs as the electrode materials in supercapacitors, in this chapter we introduce the potential of VGs in other energy storage and conversion devices, specifically, rechargeable batteries, fuel cells, and dye-sensitized solar cells (DSSCs). Without the use of any binder and conductive additives, VGs grown on current collectors (e.g., metal foils) can directly serve as the anode in lithium-ion batteries (LIBs) and the electrode in vanadium redox flow batteries (VRFBs), which can significantly simplify the battery manufacturing process. More important, batteries using VG electrodes have shown improved performance. For energy conversion applications, VGs can be used in fuel cells and DSSCs. The unique features of VGs make them an ideal catalyst support for anode reactions in a fuel cell. In addition to working as the catalyst support, VGs with a controlled number of oxygen functional groups can work as catalysts themselves for DSSCs.

Keywords Lithium-ion battery · Vanadium redox flow batteries · Fuel cell · Dye-sensitized solar cell

The rapidly increasing global energy consumption leads to large emissions of greenhouse gases and dramatic climate change. To meet the growing energy demand while avoiding environmental deterioration, it is imperative to develop sustainable solutions for clean and renewable energy. Thus, tremendous endeavors have been made in the research of advanced technologies for more efficient energy storage and conversion devices.

Part of this chapter was adapted from our review article "Emerging Energy and Environmental Applications of Vertically-Oriented Graphenes," Chemical Society Reviews, 2015 (DOI: 10.1039/C4CS00352G)—Reproduced by permission of The Royal Society of Chemistry.

J. Chen et al., *Vertically-Oriented Graphene*, DOI 10.1007/978-3-319-15302-5_8

Lithium-ion batteries (LIBs) and vanadium redox flow batteries (VRFBs) have attracted considerable interest due to their high energy density and low maintenance (low self-discharge and no periodic discharge needed). They can find wide applications ranging from renewable energy, portable electronic devices, to electric vehicles and grid storage [1]. Solar cells have been considered as a technical and economical alternative to Si-based p–n junction photovoltaic devices [2, 3]. Fuel cells are devices that convert chemical energy from fuels, e.g., hydrogen, methanol, ethanol into electricity through chemical reactions with oxidizing agents, e.g., oxygen [4–6]. Vertically-oriented graphenes (VGs) with good electrical conductivity, robust binding with the substrate, and high chemical tolerance in electrolytes, present a great promise for applications in the above-mentioned energy storage/conversion devices [7].

8.1 VG-Based Active Materials for Lithium-Ion Batteries

As one of the most common rechargeable batteries, LIBs store charges through the reversible insertion/extraction of lithium ions between redox-active host materials (i.e., anode and cathode) [8]. During charge/discharge processes, lithium ions are inserted into/extracted from the anode materials. In the initial process, a stable passivation layer (known as solid-electrolyte interphase, SEI) is formed at the anode-electrolyte interface, which is permeable for Li^+ ions but not for the electrolyte, leading to a barrier to prevent further degradation. Meanwhile, it should be noted that a high current drawn from an electrode material can cause a drop in the potential due to its internal resistance and rate limitation, an effect known as polarization [8]. This effect decreases the available power and energy from the cell. Consequently, high reversible lithium storage capability and good cycling stability are desirable for anode materials.

VGs directly grown on bare [9] or graphene-coated [10] metal foil current collectors are promising anode materials for LIBs. VGs with an exposed graphene surface and edge planes can provide numerous sites for the capture of Li ions. Meanwhile, the open intersheet channels, vertical alignment, and excellent electrical connection to the substrate of VGs can significantly reduce the transport resistance of Li ions, the intrinsic resistance within the anode materials, and the contact resistance between the anode materials and the current collector, respectively [7]. This is why LIBs employing VGs as anode materials present a very high reversible lithium storage capacity and excellent cycling stability [9, 10].

Xiao et al. reported a VG-based LIB with a high rate capability [9]. A mixture of acetylene, hydrogen, and argon was used as precursors to fabricate VGs in a microwave plasma-enhanced chemical vapor deposition (MW-PECVD) system. Nickel foil was chosen as the substrate and the system was maintained at 40 torr and 800 °C for deposition. The as-obtained VGs directly served as an electrode, exhibiting high Li-ion insertion/extraction performance. Figure 8.1a shows the capacity retention of a VG electrode as a function of the cycle number. It was

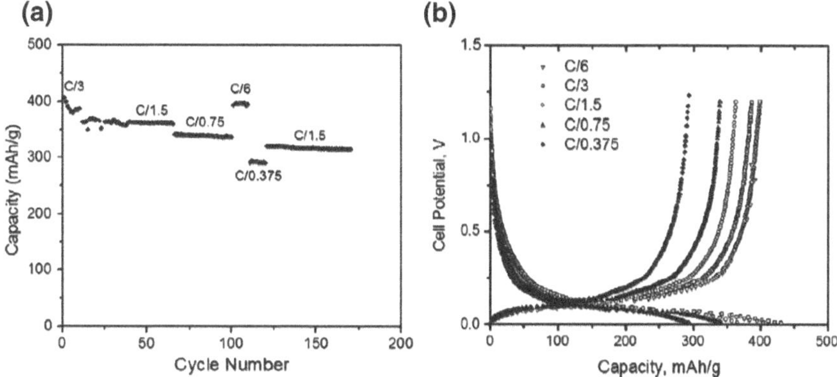

Fig. 8.1 **a** Capacity retention of the VG electrode as a function of cycle number [9]. **b** Charge–discharge profiles of the VG electrode as a function of rate. Reprinted with permission from [9]. Copyright 2011 Elsevier. Even at a high rate of C/0.375, the plateau features on the profiles can still be seen. 1C rate means that at the discharge current, the battery can discharge completely in 1 h. For example, for a battery with a capacity of 10 Ah, 1C equates to a discharge current of 10 A

found that a VG electrode could be reversibly cycled between 0.01 and 1.2 V for over 150 cycles with limited capacity loss. Figure 8.1b presents charge–discharge profiles of the VG electrode as a function of rate (from C/6 to C/0.375). Notably, even at a high rate of C/0.375, a reversible capacity of 280 mAh/g or ~74 % of the low rate capacity was retained, suggesting the good rate capability of the VG-based anode.

Kim et al. proposed a novel graphene-based hybrid nanostructure as the anode material of LIBs [10]. VGs were grown on a CVD-grown graphene film using a DC-PECVD method under atmospheric pressure. The morphology of VGs was characterized by SEM and TEM. As shown in Fig. 8.2a, spherical granule-like VGs were uniformly grown on the surface of the substrate. Since VGs were directly connected to the graphene film on the current collector (Fig. 8.2b), this surface-bound graphene network was expected to offer low intrinsic resistance at the electrode interface and significantly promote the electron transfer. The hybrid

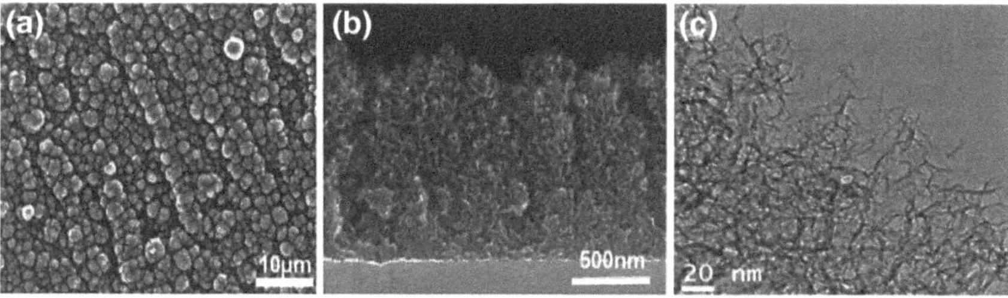

Fig. 8.2 SEM images of *top view* (**a**) and *side view* (**b**) of VG on a graphene film on a Cu foil; granule-like VG is uniformly grown on the graphene film (**c**). TEM image of VG. Reprinted with permission from [10]. Copyright 2012 Royal Society of Chemistry

graphene–VG nanostructure presented nanosheets with a very small lateral dimension of <5 nm (Fig. 8.2c), which could potentially provide plenty of lithium storage sites.

Due to the advantages such as its unique morphology, high electrical conductivity, large interfacial surface area, and high porosity, the graphene–VG hybrid nanostructure delivered promising electrochemical performance [10]. As shown in Fig. 8.3a, the graphene–VG hybrid structure showed a specific capacity of 461 mAh/g at a current density of C/5 (10 % irreversible capacity loss after 100 cycles), which is much higher than that of the theoretical specific capacity of

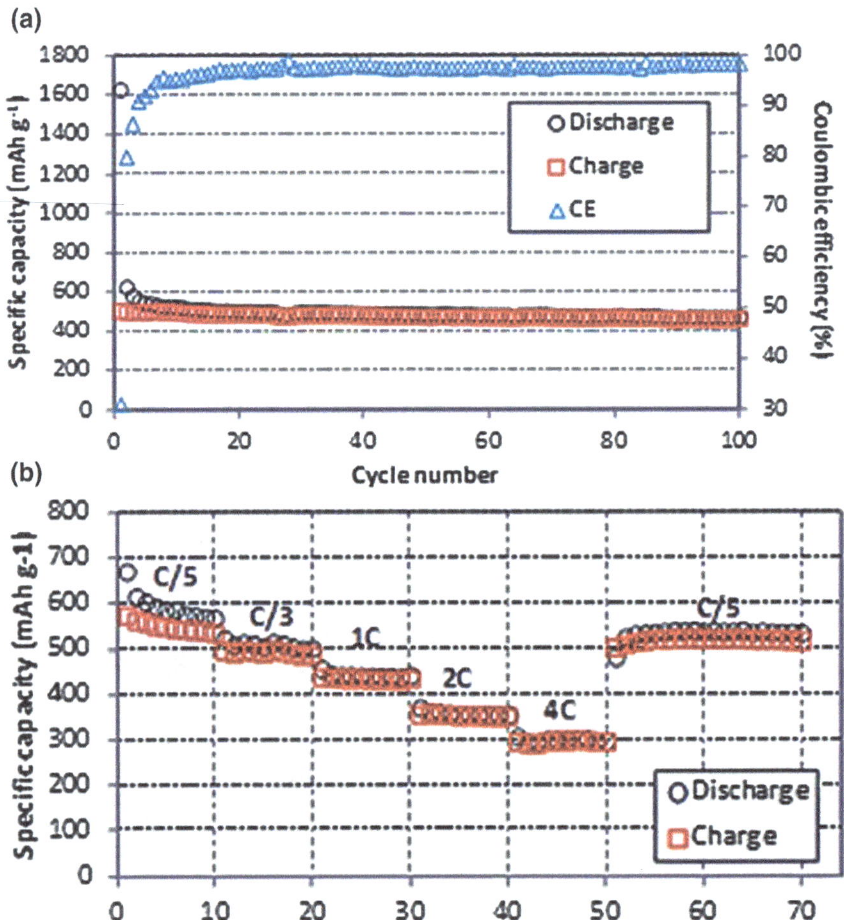

Fig. 8.3 **a** Cycle performance (charge, discharge, and coulombic efficiency or CE) between 0.01 and 2.0 V at a rate of C/5 (current density of 50 mA/g) for VG with a graphene underlayer. **b** Rate performance of VG on the graphene film at various current densities; the coin cell was cycled at C/26 in the first 3 cycles and subsequently cycled at higher current rates. Reprinted with permission from [10]. Copyright 2012 Royal Society of Chemistry

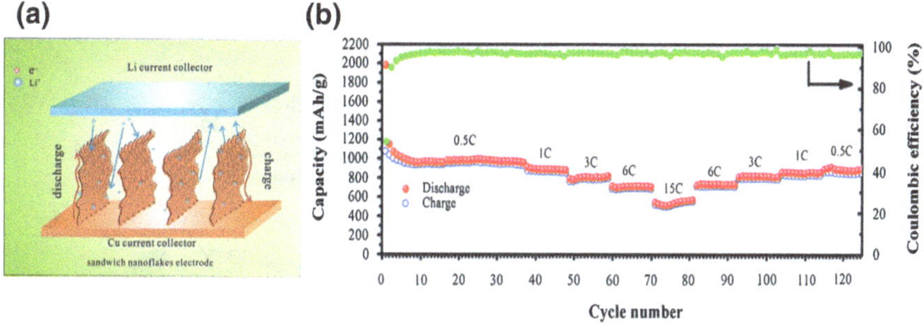

Fig. 8.4 **a** Schematic of the Li-ion diffusion mechanism in VG@GeOx sandwich nanoflakes-based electrode. **b** Performance of the VG@GeOx electrode (70 % GeOx) at the rate of 0.5C, 1C, 3C, 6C, and 15C. Reprinted with permission from [13]. Copyright 2013 Elsevier

graphite (372 mAh/g, corresponding to a stoichiometry of LiC_6). A good specific capacity of 297 mAh/g was retained even at a high rate of 4C (Fig. 8.3b). When lowering back to C/5, a reversible capacity of 518 mAh/g was obtained, indicating the high cycle stability of this novel VG-based material.

Moreover, hybrid nanostructures combining VGs and lithium alloy-forming materials were also developed as a high-performance anode of LIBs. As is well known, Sn, SnO_2, Si, and GeO_x have relatively high theoretical specific capacities of 994, 781, 4,200, and 1,000 mAh/g, respectively [8, 11–13]. Despite the high theoretical capacities, these alloys usually suffer from their poor electrical conductivity and poor cycle stability for LIB applications. In this regard, Jin et al. demonstrated that the combination of VGs and GeO_x (lithium alloying material) can realize improved lithium capture [13]. VGs were synthesized by a MW-PECVD system. Then, GeO_x was deposited on VGs via a thermal Ge/Sn coevaporation method. In such a VG@GeO_x sandwich-nanostructured anode, as shown in Fig. 8.4a, [13], VGs can work as fast electron transport channels and ensure smooth lithium diffusion pathways. The VG@GeO_x anode presented a stable capacity of 1008 mAh/g at 0.5C (retention of 96 % capacity after 100 cycles), a capacity of 545 mAh/g at a high rate of 15C, and a capacity retention of 92 % when the rate recovered to 0.5C (see Fig. 8.4b). These characteristics are considered very competitive compared with those of other lithium alloying material-based LIBs [13].

8.2 VG-Based Active Materials for Vanadium Redox Flow Batteries

As schematically shown in Fig. 8.5, VRFBs normally employ vanadium ions in different oxidation states to store chemical energy. Unlike other types of batteries, energy is stored in two liquid electrolytic solutions in VRFBs. As a consequence, VRFBs exhibit advantages such as a high energy efficiency, a long life, a flexible

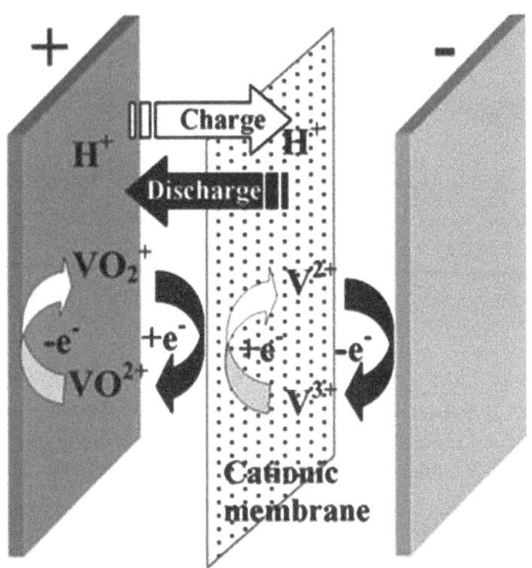

Fig. 8.5 Schematic of a vanadium redox flow battery. Reprinted with permission from [14]. Copyright 2006 Elsevier

design, no cross-contamination, and a low maintenance cost [14]. The electrodes in VRFBs are not directly involved in the storage of energy. They are normally used as the support for vanadium reactions, and thus a high surface area with active sites and a high electrical conductivity are desirable features of VRFB electrodes.

The potential advantages of VGs as the electrode of VRFBs mainly include the following [7]:

(i) The dense exposed edge planes of VGs with oxygen-based functional groups act as active sites for vanadium reactions (Fig. 8.6).
(ii) The interconnected VG networks perpendicular to the substrate can facilitate the charge transfer.

Gonzalez et al. demonstrated that VG films were suitable as the electrodes in VRFBs [15]. With changing Ar flow rates (700, 1,050, and 1,600 sccm) during the process, VG films with different characteristics were obtained with an RF-PECVD reactor. These samples were then used as electrode materials in a positive half-cell of VRFB. At a scan rate of 50 mV/s, the VG-based electrode exhibited the highest peak current densities of 15.25 and 14.32 mA/cm^2 for the anodic and cathodic processes, respectively. In addition, as for repetitive CV tests (100 scans for each voltage), there was no significant change in peak current densities or potential values, indicating the long-term stability of the VG electrode.

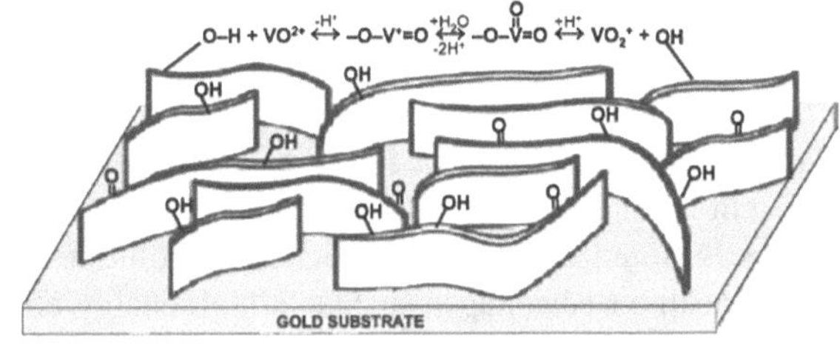

Fig. 8.6 Proposed mechanism for the $V^{4+/5+}$ redox reactions on VG thin films. Reprinted with permission from [15]. Copyright 2012 Elsevier

8.3 VG-Based Active Materials for Fuel Cells

Fuel cells can convert chemical energy from fuels (e.g., methanol and hydrogen) into electricity through a chemical reaction with oxygen or other oxidizing agents. In a typical fuel cell system, chemical reactions occur at the interfaces of three different segments, i.e., anode, electrolyte, and cathode. At the anode, a catalyst is normally employed to assist the oxidation of fuels and thus the catalytic behavior of the catalyst strongly influences the fuel cell performance.

The unique features of VGs make them ideal catalyst supports for such anode reactions [7]. The non-agglomerated morphology and open channels of VG can enhance the deposition and dispersion of catalysts. Bo et al. previously reported the electrocatalytic performance of VGs supported Pt–Ru bimetallic catalysts toward methanol oxidation reaction (MOR) [16]. VGs were directly grown on carbon paper (CP) via a MW-PECVD method. Then, a repeated pulse potential approach was applied for the coelectrodeposition of Pt–Ru bimetallic nanoparticles on CP and VG-coated CP from a typical electrolyte composition, i.e., $[H_2PtCl_6]:[RuCl_3] = 1:1$. As a result, VG-coated CP exhibits a ~3.5 times higher catalyst mass loading and a ~50 % smaller nanoparticle size than the pristine CP (Fig. 8.7a, b). The Tafel plots (Fig. 8.7c) were composed of two nearly linear regions that intersect at approximately 0.45 V, indicating a change in the rate-determining steps. In the region below 0.45 V, the Tafel slopes of Pt/VG and Pt–Ru/VG were 119 mV/dec ($\alpha = 0.50$) and 194 mV/dec ($\alpha = 0.31$), respectively. Hence, it was evident that the electrodes could achieve a fast methanol dehydrogenation, compared with the theoretical prediction of 119 mV/dec for a one-electron transfer reaction as a rate-determining step [17]. As shown in Fig. 8.7c, d, Pt–Ru/VG could realize a faster methanol dehydrogenation than Pt/VG. In addition, as shown in Fig. 8.8d, VG support presented a significantly higher current density of 339.2 mA/mg than by using pristine CP (116.3 mA/mg) and commercial carbon black Vulcan XC-72 (192.8 mA/mg).

Moreover, fast electron transport between the reaction sites and the current collector can be realized along the in-plane direction of the graphene nanosheets [7]. Specifically for the MOR in direct-methanol fuel cells, Pt nanoparticles supported by VGs exhibited more improved electrocatalytic performance in activity and stability than the counterpart employing vertically aligned carbon nanofibers (VACNFs) [18]. Both VGs and VACNFs were directly grown on CP via an inductively coupled plasma-enhanced vapor deposition (ICP-PECVD). To fabricate Pt/VG and Pt/VACNF electrodes, Pt particles with a loading of 0.025 mg/cm^2 were deposited on the surface of VGs and VACNFs through a radio-frequency magnetron sputtering system. As shown in Fig. 8.8a, the methanol oxidation of JM 40 % Pt/C, Pt/VACNFs, and Pt/VGs was evaluated by CV measurements. It can be seen that the Pt/VGs exhibited the higher normalized current density (1,524 mA/(cm^2 mg) Pt) of methanol oxidation, compared with Pt/VACNFs (1,119 mA/(cm^2 mg) Pt) and JM 40 % Pt/C (548 mA/(cm^2 mg) Pt), indicating a superior electrocatalytic activity toward methanol oxidation. Also the Pt/VG

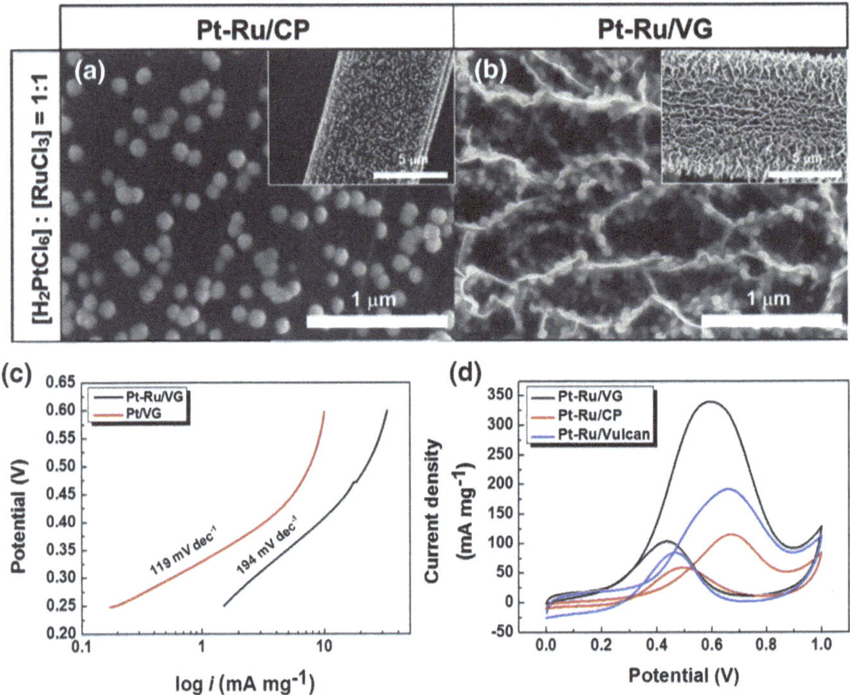

Fig. 8.7 SEM images of Pt–Ru/CP and Pt–Ru/VG obtained from $[H_2PtCl_6]$:$[RuCl_3]$ = 1:1, **a** and **b** average catalyst diameter: 111.1 ± 2.8 nm and 51.2 ± 1.8 nm. **c** Tafel curves of Pt/VG and Pt–Ru/VG. **d** CV curves of Pt–Ru/VG, Pt–Ru/CP, and Pt–Ru/Vulcan in a N_2-saturated solution of 0.5 M H_2SO_4 + 1 M CH_3OH at a scan rate of 50 mV/s. Reprinted with permission from [16]. Copyright 2015 Elsevier

electrode expressed a higher value of 1.2 for the ratio of the forward anodic peak current density (I_f) to the reverse anodic peak current density (I_b) than that of the Pt/VACNF electrode (1.1), indicating improved carbon monoxide (CO) tolerance. These characteristics are considered to be associated with the unique structure of VG, which results in a faster electron transport and a shorter electron transport path compared with the Pt/VACNFs electrode, as shown in Fig. 8.8b, c.

In addition, transition metal dichalcogenides supported by VGs have also been demonstrated as effective catalysts for fuel generation. Electrolysis of water is one of the major sources to produce hydrogen. In this case, VGs were grown on a graphite disk by a PECVD method, followed by CVD to grow $MoSe_2$ nanosheets on VG. After the growth, the $MoSe_2$/VG hybrids were used as an electrocatalyst for a hydrogen evolution reaction (HER), showing a greatly improved catalytic activity for the HER compared with the bare $MoSe_2$ nanosheets (Fig. 8.9a) [19]. A remarkable positive shift of the onset potential was found in the polarization curves of the catalysts, confirming that the HER was catalyzed at a lower overpotential with the $MoSe_2$/VG catalysts (Fig. 8.9b).

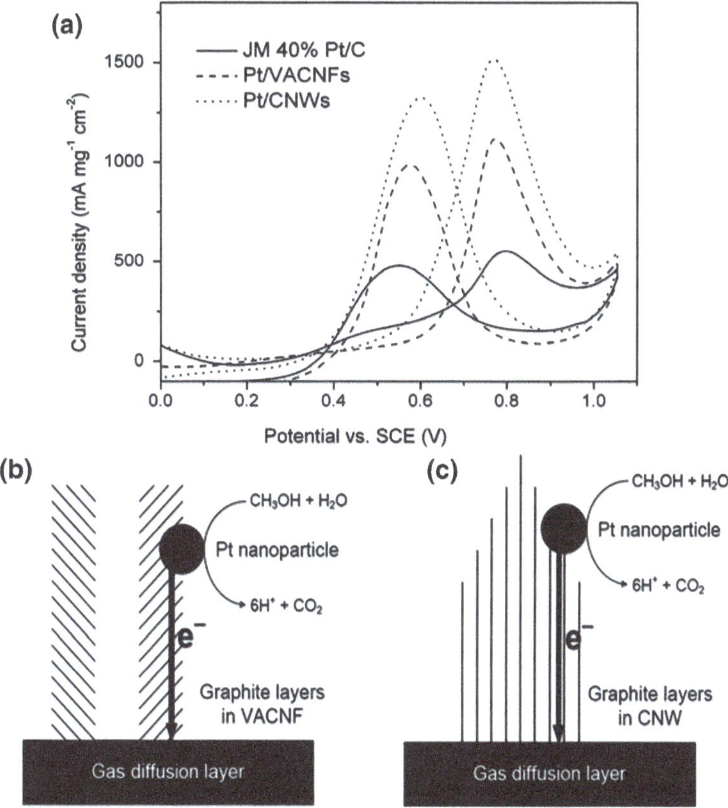

Fig. 8.8 **a** CVs of JM 40 wt% Pt/C, Pt/VACNFs and Pt/VGs ('VG' was labeled as 'CNW' in the figure) electrodes in N_2-saturated 1 M H_2SO_4 + 2 M MeOH at a scan rate of 0.05 V/s. Current densities are normalized with respect to the Pt loading. **b** Reaction and electron transfer schemes of the Pt/VACNF electrode. **c** Reaction and electron transfer schemes of Pt/VG electrode. Reprinted with permission from [18]. Copyright 2012 Elsevier

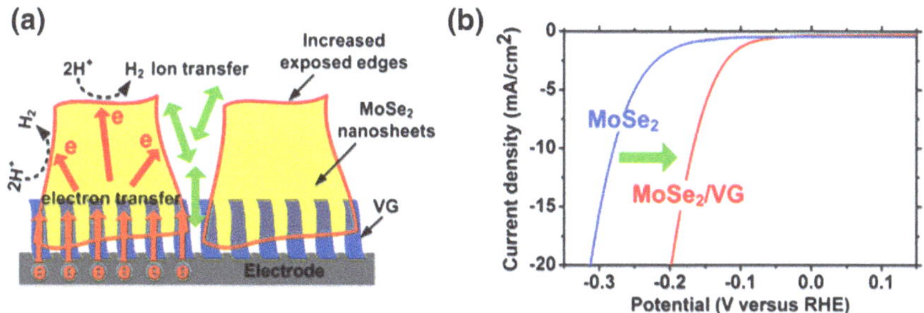

Fig. 8.9 **a** Schematic of HER and electron transport between the perpendicularly oriented $MoSe_2$ nanosheets, VG, and the electrode. **b** Improved hydrogen evolution catalytic activity of the $MoSe_2$ nanosheets with a VG support. The electron transfer at the electrode interface is greatly promoted through the highly conductive VG, which smoothly bridges $MoSe_2$ nanosheets and the current collector due to the in situ growth of both VG and $MoSe_2$ nanosheets. Reprinted with permission from [19]. Copyright 2014 Wiley

8.4 VG-Based Active Materials for Solar Cells

Solar cells. Besides working as the catalyst support, VGs with a controlled number of oxygen functional groups can work as catalysts themselves for dye-sensitized solar cells (DSSCs). DSSCs use light-absorbing dye molecules to generate electricity from sunlight. In a typical DSSC operation, I_3- is reduced at the counterelectrode for dye regeneration. Conventional Pt catalysts possess a high catalytic activity for I_3- reduction but suffer from a high cost. Because of the large surface area and a low cost, VG with oxygen functional groups (catalytic sites) was demonstrated as a promising substitute for Pt in the counterelectrode of DSSCs [20, 21]. As illustrated in Fig. 8.10a, the charge was mainly transferred from the functional groups into the iodide molecules [20]. Hence, VGs with a large number of oxygen functional groups can contribute greatly to catalytic performance (Fig. 8.10b). The DSSC with a VG electrode exhibited an improved power conversion efficiency (PCE) of the cell from 4.68 % (for DSSCs with a Pt electrode) to 5.36 % (Fig. 8.10c), with a charge transfer resistance of about 1 % of the Pt electrode.

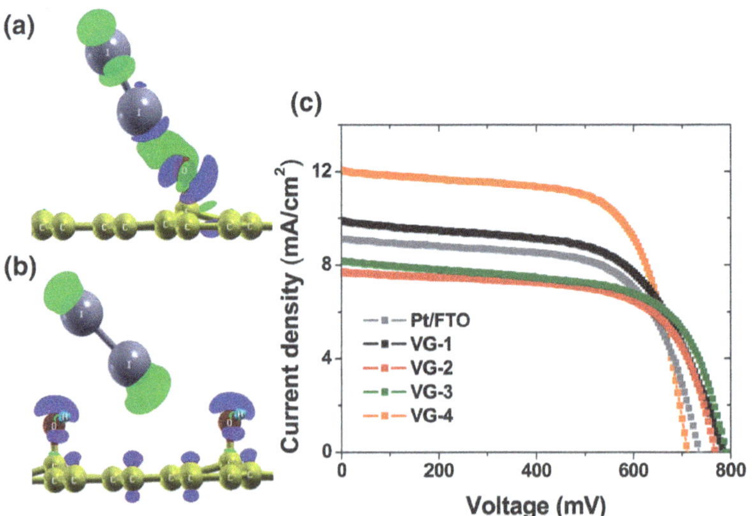

Fig. 8.10 Side views of contour plots of charge transfer between molecular I_2 and graphene oxidized with functional groups, **a** epoxy with charge transfer constant 0.00250 e and **b** hydroxyl with charge transfer constant 0.00125 e. The *green* and the *purple color* represent charge accumulation and depletion zone, respectively. **c** Current density–voltage characteristics of DSSCs with platinized fluorine-doped tin oxide (FTO) and VG counterelectrodes. The illumination intensity is one sun, AM 1.5 G, 100 mW/cm^2. Reprinted with permission from [20]. Copyright 2012 Royal Society of Chemistry

8.5 Summary

VG can be used for developing high-performance electrochemical energy storage and conversion devices, considering its outstanding electrical conductivity, robust binding with the substrate, and high chemical tolerance in electrolytes. After discussing the application of VG in supercapacitors in Chap. 7, we have introduced in this chapter the potential of VG in several other electrochemical energy devices: specifically, batteries, fuel cells, and (DSSCs). LIBs and VRFBs using VG electrodes have shown improved performance (e.g., high reversible lithium storage capacity and excellent cycling stability). In addition, the use of VG grown directly on current collectors can greatly simplify the battery manufacturing process, since no binder or conductive additives are needed. For energy conversion applications, VG has also shown potential as an ideal catalyst support for anode reactions in a fuel cell and as the catalyst itself for DSSCs.

References

1. Kumagai, J. (2012). A battery as big as the grid. *Spectrum IEEE, 49*(1), 45–46.
2. Grätzel, M. (2003). Dye-sensitized solar cells. *Journal of Photochemistry and Photobiology C-Photochemistry Reviews, 4*(2), 145–153.
3. O'Regan, B., & Gratzel, M. (1991). A low-cost, high-efficiency solar cell based on dye-sensitized colloidal TiO_2 films. *Nature, 353*(6346), 737–740.
4. Logan, B. E., Hamelers, B., Rozendal, R. A., Schrorder, U., Keller, J., Freguia, S., et al. (2006). Microbial fuel cells: Methodology and technology. *Environmental Science and Technology, 40*(17), 5181–5192.
5. Steele, B. C. H., & Heinzel, A. (2001). Materials for fuel-cell technologies. *Nature, 414*(6861), 345–352.
6. Winter, M., & Brodd, R. J. (2004). What are batteries, fuel cells, and supercapacitors? *Chemical Reviews, 104*(10), 4245–4269.
7. Bo, Z., Mao, S., Han, Z. J., Cen, K., Chen, J., & Ostrikov, K. (2015). Emerging energy and environmental applications of vertically-oriented graphenes. *Chemical Society Reviews*, doi: 10.1039/C4CS00352G.
8. Roberts, A. D., Li, X., & Zhang, H. (2014). Porous carbon spheres and monoliths: morphology control, pore size tuning and their applications as Li-ion battery anode materials. *Chemical Society Reviews, 43*(13), 4341–4356.
9. Xiao, X., Liu, P., Wang, J. S., Verbrugge, M. W., Balogh, M. P. (2011). Vertically aligh-ned graphene electrode for lithium ion battery with high rate capability. *Electrochemistry Communication, 13*, 209–212.
10. Kim, H., Wen, Z., Yu, K., Mao, O., & Chen, J. (2012). Straightforward fabrication of a highly branched graphene nanosheet array for a Li-ion battery anode. *Journal of Materials Chemistry, 22*(31), 15514–15518.
11. Winter, M., Besenhard, J. O., Spahr, M. E., & Novák, P. (1998). Insertion electrode materials for rechargeable lithium batteries. *Advanced Materials, 10*(10), 725–763.
12. Ji, L., Lin, Z., Alcoutlabi, M., & Zhang, X. (2011). Recent developments in nanostructured anode materials for rechargeable lithium-ion batteries. *Energy and Environmental Science, 4*(8), 2682–2699.

13. Jin, S. X., Li, N., Cui, H., & Wang, C. X. (2013). Growth of the vertically aligned graphene@amorphous GeOx sandwich nanoflakes and excellent Li storage properties. *Nano Energy, 2*(6), 1128–1136.
14. de León, Ponce, Frías-Ferrer, C. A., González-García, J., Szánto, D. A., & Walsh, F. C. (2006). Redox flow cells for energy conversion. *Journal of Power Sources, 160*(1), 716–732.
15. Gonzalez, Z., Vizireanu, S., Dinescu, G., Blanco, C., & Santamaria, R. (2012). Carbon nanowalls thin films as nanostructured electrode materials in vanadium redox flow batteries. *Nano Energy, 1*(6), 833–839.
16. Bo, Z., Hu, D., Kong, J., Yan, J., & Cen, K. (2015). Performance of vertically oriented graphene supported platinum–ruthenium bimetallic catalyst for methanol oxidation. *Journal of Power Sources, 273*, 530–537.
17. Velázquez-Palenzuela, A., Centellas, F., Garrido, J. A., Arias, C., Rodríguez, R. M., Brillas, E., & Cabot, P.-L. (2011). Kinetic analysis of carbon monoxide and methanol oxidation on high performance carbon-supported Pt–Ru electrocatalyst for direct methanol fuel cells. *Journal of Power Sources, 196*(7), 3503–3512.
18. Zhang, C. X., Hu, J., Wang, X. K., Zhang, X. D., Toyoda, H., Nagatsu, M., & Meng, Y. D. (2012). High performance of carbon nanowall supported Pt catalyst for methanol electro-oxidation. *Carbon, 50*(10), 3731–3738.
19. Mao, S., Wen, Z., Ci, S., Guo, X., Ostrikov, K., & Chen, J. (2015). Perpendicularly oriented $MoSe_2$/graphene nanosheets as advanced electrocatalysts for hydrogen evolution. *Small, 11*(4), 414–419.
20. Yu, K., Wen, Z., Pu, H., Lu, G., Bo, Z., Kim, H., et al. (2013). Hierarchical vertically oriented graphene as a catalytic counter electrode in dye-sensitized solar cells. *Journal of Materials Chemistry A, 1*(2), 188–193.
21. Yang, C., Bi, H., Wan, D., Huang, F., Xie, X., & Jiang, M. (2013). Direct PECVD growth of vertically erected graphene walls on dielectric substrates as excellent multifunctional electrodes. *Journal of Materials Chemistry A, 1*(3), 770–775.

Chapter 9
Conclusions and Outlook

Abstract Vertically-oriented graphene (VG) is an attractive material with outstanding properties and great potential for various applications. After discussing in previous chapters VG's properties, its synthesis using plasma-enhanced chemical vapor deposition (PECVD), and its various applications, in this chapter we provide some conclusions and an outlook for this amazing material. PECVD has been successfully demonstrated for VG growth on various substrates of different materials and geometries. A series of key parameters (e.g., plasma source, feedstock gas, temperature and pressure, electric field, and localized current density) are identified in PECVD that affect the plasma behavior and thus the VG growth. However, there is still a need to build a unified theory to unveil the VG growth mechanism and to provide guidance for optimum growth conditions using different plasma sources. For broad VG applications, further development of VG growth techniques is required to solve the following three challenges: morphology-/structure-controlled growth, large-scale production, and low-temperature growth. The unique structures and intrinsic properties of VG greatly expand the use of conventional graphene sheets, and more applications in diverse fields are anticipated in the near future. VG's outstanding performance will encourage and inspire additional studies on the exploration and use of other vertically-oriented 2-D nanostructures.

Keywords Graphene · Growth mechanism · Growth rate · Morphology · Orientation · Plasma-enhanced chemical vapor deposition · Vertically-oriented graphene

---

Part of this chapter was adapted from our review article "Emerging Energy and Environmental Applications of Vertically-Oriented Graphenes," Chemical Society Reviews, 2015 (DOI: 10.1039/C4CS00352G)—Reproduced by permission of The Royal Society of Chemistry.

J. Chen et al., *Vertically-Oriented Graphene*, DOI 10.1007/978-3-319-15302-5_9

9.1 Conclusions

PECVD employing a variety of plasma sources such as MW, RF, and DC discharges with different reactor configurations has been successfully demonstrated for vertically-oriented graphene (VG) growth. The VG growth can be realized not only on the full area or selected area of convenient planar substrates, but also on cylindrical and CNT substrates. While extensive studies have been conducted to reveal the influence of various key parameters on plasma behavior and VG growth performance, such as feedstock gas type and proportion, temperature and pressure, and electric field and localized current density, there is still a need to build a unified theory to unveil the growth mechanism (e.g., the detailed growth process, key species influencing VG morphology and structure, and the critical factor to aligned growth) and to provide guidance for optimum growth conditions (e.g., the selection of a-C etchants, the proper gas proportion, and the optimized growth temperature and pressure for specific applications) for different plasma sources. Further understanding of several key issues on controlled VG growth and plasma chemistry is warranted for fast, large-scale production of high-quality VG and its successful applications.

In contrast to the conventional horizontal graphene stacks obtained by various chemical or physical routes, VGs show many unique features such as vertical orientation, non-agglomerated morphology, high surface-to-volume ratio, and exposed sharp edges. Combined with the inherent excellent electrical, chemical, and mechanical properties of graphenes, VGs show great potential and several advantages in conventional and emerging applications ranging from field emission, sensing (biosensors and gas sensors), green corona discharges, solar cells, to energy storage devices (supercapacitors and rechargeable batteries).

9.2 Challenges and Outlook

The advantages and unique properties of VG have been extensively reported. Future VG applications call for further understanding and improvement in VG production in the following three directions, i.e., morphology-/structure-controlled growth, massive production, and low-temperature growth.

The VG morphology and structure strongly influence the application performance; therefore, the growth of VG in a controlled manner is required. Better understanding of the VG growth mechanism could help to provide information and guidance for the controlled growth. The first step is to identify the key parameter determining the growth orientation. Different from the CVD of graphene, whose growth mechanism has been well proposed, the initial nucleation and subsequent evolution of VG are yet to be convincingly interpreted. Although many growth mechanisms have been proposed to understand the vertical growth in PECVD synthesis of VG, details in the growth preference and underlying reasons still need more investigation. For example, since a catalyst was commonly not a necessity in VG growth, much research has been conducted to focus on its initial nucleation stage and the subsequent vertical growth. It is suggested that the initially nucleated graphene sheets were randomly oriented; however, those standing almost vertically on the substrate grow faster due to the growth rate difference between the directions along the expanding graphene sheets and along the stacking, the preference in radical diffusion, and the overshadow effect. As demonstrated in a VG growth model, the carbon diffusion and the electric field are two critical factors that determine the VG growth. The high surface diffusion rates, caused by the large difference between the surface adsorption energy (~0.13 eV) and the surface diffusion energy (~1.7 eV) for a carbon atom (comes from the growth species) on the graphene surface, could help the migration of carbon atoms along the graphene surface leading to a vertical growth. Meanwhile, the introduction of a proper electric field was believed to help the alignment of VG. Another important issue is the careful selection of growth precursors and the feed gas proportion.

In addition to the morphology-/structure-controlled growth, the growth of VG at a high growth rate on large areas remains a critical prerequisite for industrial applications. Compared with graphene prepared by wet-chemical methods, i.e., the synthesis of GO by the modified Hummer's method and the subsequent reduction of GO, PECVD synthesis can result in the oriented growth of graphene nanosheets with non-agglomerated morphology; however, its disadvantages include a low yield and a relatively low growth rate, which are potential inherent drawbacks of PECVD techniques (not only for the growth of VG). It should be noted that an important factor leading to these disadvantages is the operating pressure. As mentioned, the feedstock gas flow rate and the plasma energy are two competing factors: most PECVD processes are conducted at a low pressure to achieve a relatively long mean free path of electrons while the massive production of VG calls for a large volume of gas input. From this point of view, the atmospheric growth with the elevated plasma energy could be used to solve this problem.

Finally, the growth of high-quality VG at a lower temperature would make it more practical in many aspects. For example, fabrication of VG on high-performance display glass needs a growth temperature lower than 660 °C. Furthermore, VG would become compatible with flexible substrates (e.g., plastics) by further lowering the growth temperature, opening new applications of VG in the area of flexible electronic devices.

In addition to the challenges in VG synthesis, there is no doubt that the potential of VGs in a wide range of energy and environmental applications has not been fully exploited. For example, the specific capacitance of VG-based supercapacitor electrodes still requires further improvement. The FET biosensors and gas sensors have limitations arising from the intrinsic electrical properties of VGs, i.e., very narrow bandgap and low on–off current ratio. In addition, for practical purposes, critical factors such as reliability, selectivity, and robustness of sensors still require comprehensive assessments according to industrial standards. A number of further studies are warranted to better use the VGs' unique features. For real-world applications, better understanding of the functional performance of VGs is essential for the optimum material design/fabrication, performance optimization, as well as the development of scale-up processes of technological relevance. Examples of such mechanisms include mass transfer and charge storage at the interface of electrolytes and VGs for supercapacitors, and electron transport during the interactions of VGs with bio- or gas molecules. Moreover, surface engineering and doping of VGs could also enhance the VG device performance.

The unique structures and intrinsic properties of VGs greatly expand the use of conventional graphene sheets, and more applications in diverse fields are anticipated in the near future. Finally, VG's outstanding performance will encourage and inspire additional studies on the exploration and use of other vertically-oriented 2-D nanostructures.

9.3 Summary

In this chapter, we have provided some conclusions and an outlook for PECVD synthesis and applications of VG, which is an amazing material with outstanding properties and great potential for various applications. PECVD processes have been widely and successfully used for VG growth on various substrates. Important process factors in PECVD include the plasma source, the feedstock gas, the temperature and pressure, the electric field, and the localized current density, which affect the plasma behavior and thus the VG growth. However, it is still imperative for researchers in the VG field to build a unified theory to unveil the VG growth mechanism and to provide guidance for the optimum growth conditions using different plasma sources. We have also identified some challenges that need to be addressed for the further development of VG growth techniques and for widespread applications of VG. Because of its unique structures and intrinsic properties, with further efforts, the future for VG is bright and we will soon see its use in prototypes or even market-ready devices/systems in diverse areas.

Index

A
Amorphous carbon, 41
Anisotropic growth effects, 20, 21
Argon, 43, 45
Atmospheric pressure, 56, 63

B
Ballistic electron mobility, 12

C
Carbon, 1–3
Carbon nanotube, 61, 62
Carbon source, 35–37, 42, 43, 45
Corona discharge, 74–76

D
Direct current discharge, 23, 28
Dye-sensitized solar cell, 106, 107

E
Electric double-layer capacitor, 80–85, 88, 89, 94
Electric field, 20–22, 25–27, 29, 56, 59, 62
Electrochemical biosensor, 70, 71
Electronic biosensor, 68
Electron mobility, 2
Environmental application, 76
Etchant, 41–43, 45–47

F
Feedstock gas, 35, 42, 46, 47, 51
Fuel cell, 98, 103, 107

G
Gas proportion, 47
Gas sensor, 72–74
Graphene, 2, 3, 42, 56, 59, 61, 63, 110, 111
Growth mechanism, 110, 112
Growth rate, 110, 111

H
Hybridization, 1, 12

I
Internal stress, 20, 21

L
Lithium-ion battery, 97

M
Microwave plasma, 24
Morphology, 110, 111

N
Nitrogen, 42, 46

J. Chen et al., *Vertically-Oriented Graphene*, DOI 10.1007/978-3-319-15302-5

O

Orientation, 13, 14, 39, 59, 110

P

Patterned growth, 58, 59
Plasma-enhanced chemical vapor deposition (PECVD), 3, 4, 15, 35, 39, 42, 43, 45–47, 49, 51, 55–61, 63, 109–112
Precursor, 35, 36, 38, 39, 42
Pressure, 55–57, 63
Pseudo-capacitor, 80, 88–90, 94

R

Radio frequency plasma, 27
Raman spectroscopy, 29

S

Spatial alignment, 2
Specific surface area, 12, 13, 15, 45, 50, 51, 57, 58, 60, 61
Substrate, 57–59, 61, 63
Supercapacitor, 80, 88, 90–94

T

Temperature, 38, 39, 45, 46, 48, 49, 51
Thermal conductivity, 12

V

Vanadium redox flow batteries, 97, 101, 107
Vertically-oriented graphene (VG), 2–4, 13–16, 42, 56–63, 111

Zeitfracht Medien GmbH
Ferdinand-Jühlke-Straße 7
99095 Erfurt, Deutschland
produktsicherheit@kolibri360.de